SYMMETRY AND SELECTIVITY IN U.S. DEFENSE POLICY: A Grand Design Or A Major Mistake?

Agatha S.Y. Wong-Fraser
Center for Science & International Affairs
Harvard University

University Press of America™

University Press of America, Inc.
4720 Boston Way, Lanham, MD 20801

Printed in the United States of America

ISBN: 0-8191-1182-1 (Case)
0-8191-1183-X (Perfect)

Library of Congress Catalog Card Number: 80-5610

In gratitude to

My Mother and Father,

Sisters and Brothers,

and

Michael Fraser, my husband

Always love, always there

ACKNOWLEDGEMENTS

Special gratitude goes to my parents for their sponsorship of this research. I have also benefited from David Greenwood, Director of the Center for Defense Studies of the University of Aberdeen for his helpful discussions and guidance. I am also grateful for the companionship of my colleagues in the Center for Science and International Affairs, Harvard University, especially Albert Carnesale, Thomas Schelling and Robin Staffin, and for the supporting staff, Barbara Talhouni and Zarina Gomez-Mont. Typing of the final draft has been done by Mary Ann Spano-Wells whose skills and good cheer have made life pleasant for me.

My debt is greatest to Michael Fraser, my husband, who sustains me with lively moral and intellectual support.

TABLE OF CONTENTS

CHAPTER 1

INTRODUCTION

In March 1974 the United States' Secretary of Defense, James Schlesinger, announced certain important changes in American nuclear strategy. They have been characterized by some commentators as dramatic innovation - a major departure in the United States' strategic tradition. But others regard them as merely extensions or refinements of the McNamara doctrine, part of a continuing search for greater flexibility.

The purpose of this study is to explore the background to "the Schlesinger doctrine"; to elucidate its essence and its rationale; and to examine some of the pressures, including budgetary restraints and domestic imperatives, to which the Defense Secretary may have been responding in offering his fresh formulation. In so doing light should be shed on the question: how far was this genuine innovation, how far a development of McNamara's legacy?

In reviewing the approaches to nuclear issues in the 1950s, the 1960s and the 1970s, comparisons must be made between doctrines, postures, and also weapons-systems and the direction of research and development. In addition the overbearing political pressures that accompanied them need to be considered. This is not to say that the political process and nuclear strategy are independent factors; but it is certainly true that nuclear strategy is responsive to both the domestic political process and the international environment. The concept of threat, the concept of national security and the interpretation of alliance obligation are in turn part and parcel of the political process itself.

The subject matter of this study is examined largely in practical terms. The chief objective is to highlight trends, to assess Schlesinger's "innovation" and, whenever appropriate, discuss underlying principles. Moreover, the treatment is not cast in highly complex "systems-analysis" terms, when dealing with weapons systems. Nor is any attempt made to explore the theoretical organizational behaviouralism of the struggle between the Congress and the Defense Department over defense Issues; nor that of the various competing interests within it. No calculations or models of political relationships linking domestic pressures or public opinions and defense policy will be elaborated, although passing reference will be made to these matters as necessary. As for sources, throughout the study one will find heavy reliance on government documents, official press conferences, and articles of contemporary comment.

To avoid any misunderstanding, let me discuss, first, what I regard as the encompassing variables of strategy.

Strategy, whether it is nuclear or not, is the expression of a nation's defense policy (1) to cope with its national aspiration or destiny, (2) to pursue its national objectives and (3) to achieve its perception of national security. Doctrine, (whether it is feasible or not and whether it is a "proper" response to a particular international situation), is the abstraction of the official declaratory policy. The nature and scope of the study of strategic doctrines rests on two vital questions: What is to be studied? - the kinds of formulations, programmes, and policies to which one gives the label "strategic doctrines", and how is it to be studied? - the analytical approach used to understand formulations, programmes, and policies, once the "what" decision has been made. A first objective is to provide a language with which to talk about the evolution of American strategic doctrines

1945-1975; and to identify the factors (if any) prevalent in the evolution revolution itself.

Nation's Defense Policy

A nation's defense policy is concerned essentially with the selection of objectives and priorities, the allocation of resources and the "elaboration of the permissible parameters of government action related to the security of what are perceived to be the national interests of the state and the availability of military power'.[1] In this sense, government makes defense policy in the context of domestic as well as international pressures. The art of strategy, translated in terms of strategic doctrine, consists in choosing the most suitable means from those available to fulfill the function of a nation's defense policy. Robert Osgood has gone so far as to suggest that "military strategy must now be understood as nothing less than the overall plan for utilizing the capacity for armed coercion - in conjuction with economic, diplomatic and psychological instruments of powers - to support foreign policy most effectively by overt, covert and tacit means".[2] This is made more explicit by Henry Kissinger who argues that "it is the task of strategic doctrine to translate power into policy".[3]

The Idea of "Destiny"

The idea of "destiny" has played an important part in shaping America's attitude both toward itself and toward the rest of the world. In its purest sense the word "destiny" evokes an idealistic mission. But "destiny" is an abstraction. Yet, without characterizing the American "destiny", one falls short in understanding American strategic doctrines. "Destiny" has become an inescapable burden of world leadership which impinges upon the formulation of doctrine. That "destiny" is to defend democracy and to fight communism.

The hope that wartime alliance between the Western powers and the Soviet Union would continue after the war evaporated in face of Soviet expansionistic policies in Eastern Europe. To stop this expansion, the strategy of containment was devised. Containment, as described by George F. Kennan, was "designed to confront the Russians with unalterable counter-face at every point where they show signs of encroaching upon the interests of a peaceful and stable world"[4]. Thus described containment would allow the United States "to influence by its actions the internal developments (of Communism) both within Russia and throughout the international Communist movement". The economic and military-aid programmes of the United States, such as the Marshall Plan and the Truman Doctrine, were designed to reinforce the containment principle. The Berlin Airlift and the Korean War were examples of the use of "counter-force" in this early sense of the term.[5]

After 1945 pursuing national destiny became for the United States a mission to be done. This is exemplified in the speech that President Kennedy was to have delivered in Dallas on the day he was assassinated (November 22, 1963):

> We in this country, in this generation are - by destiny rather than choice - the watchmen on the walls of freedom. We ask, therefore, that we may be worthy of our power and responsibility, that we may exercise our strength with wisdom and restraint, and that we may achieve in our time and for all time the ancient vision of "peace on earth, good will toward men".

National Objectives

National objectives are the outcome of the selection of priorities on the allocation of resources. Seyom Brown's The Irreducible National Interest and Basic Premises About World Conditions tries to identify the "national interest" of the United States, and deduces from it specific ground rules for the implementation of foreign policy. Brown argues that debate over foreign policy in the post-1945 world has centred on contending methods of reaching ultimate goals that have remained relatively constant throughout. The best summing-up of "ultimate" national objective[6] on the military side is, perhaps, that given by Nixon: "We were determined to maintain a national defense second to none"[7]

However, other social needs make claims on the decision-making apparatus of the political system. It is here that the conflictive trail of inter-organizational relationships is likely to be found. At the same time the duplication of programmes in various organizational components concerned with national security is nearly always instrumental in shaping American strategic doctrine.

But such conflictive traits do not preclude a consensus (or compromise) in policy elaboration. In fact, a declaratory policy is the setting in which public goals and general plans for action have found expression as an interorganizational response to the converging imperatives of political demands, expressions of policy support, and feedback process from ongoing programmes. Massive retaliation (however unsound to many) was a declaration derived from budgetary restraint within the selection of national priorities. To argue, as Bernard Brodie[8] has, that interests (whether vital or nor) are the product of policy-making by fallible human beings and are, therefore, based on less than unanimous agreement about the importance of specific situations, is to forget that strategy requires both content and consensus. "Strategic policies, like statutes, are both prescriptions for future actions and ratifications of existing power relationships"[9] A strategy that sets forth no prescription for action or sets forth an unacceptable prescription is no strategy at all. Consensus is both a cost and a prerequisite in any strategic doctrine.

National Security

The phrase "national security", as Arnold Wolfers pointed out, is an "ambiguous symbol".[10] It receives much use but little analysis. The most basic element of national security is the protection of national survival. Its objective is the creation of national and international political conditions favourable to the protection - or extension - of vital national values against existing and potential adversaries. Strategic doctrine is the declaratory means to enhance national security. Security, or at least this perception of security, can only be maintained vis-a-vis the weakness of rivals or through force itself. If security is the final goal of a strategic doctrine, the effective means will be to establish a new relation of forces or to modify the old one so that potential enemies, lacking superiority, will not be tempted to take the initiative of an aggression.

The perception of national security vis-a vis the strength and intention of a potential enemy which underlies strategic doctrine is hard to define in the abstract and harder still to discuss in concrete terms. Often it is dimly perceived and understood by the policy makers (e.g. the Missile Gap) who must nevertheless base their defense policy on them.

It is common ground that the analysis of threats to national security should rest on assessment of the "capability" and "intentions" of the potential enemy. The examination of capabilities involves attention to the technical capacities and the "circumstances" which determine the policies which he could execute if he so desired. Any examination of enemy's intentions is to assess "those things which he is going to do".[11] In quasi-mathematical form: Threat Perception - Estimated Capability x Estimated intent.[12] Accordingly, strategic posture is the reflection of the analysis of threats which is the assessment of an estimation. Each perceives the other as a threat to its national security, and such perception is a function of both estimated capability and estimated intent. The various strategic doctrines attributed to different American Secretaries of Defense are the outcome of the assessment of Soviet threat. It is contended here that the shift of emphasis or any doctrinal innovation could well be explained in terms of interplay between threat perception (in this sense) and the numerous internal pressures that operate.

The progression of the nuclear balance between the superpowers from the United States' monopoly to superiority to parity complicated the discussion. The modern state is vulnerable to nuclear attack, and the sense of vulnerability is enhanced by the failure to develop effective defensive counters. War prevention has accordingly superceded "victory" as the objective of super power military dispositions. The basic problem of strategy in the nuclear age is to establish: "a relationship between a policy of deterrence and a strategy for fighting a war in case deterrence fails. The temptation of strategic doctrine is to seek to combine the advantages of every course of action: to achieve the maximum deterrence but also to do so at minimum risk".[13]

The above discussion is my attempt to disentangle some of the key factors in shaping doctrines: the sense of national "destiny" to defend democracy; the aspiration to be second to none in defense posture; the fear that the Soviet Union's military capabilities might confer advantages in the furtherance of Soviet political ambitions; and, last of all, the ever present and overarching awareness of the possibility of nuclear holocaust.

Themes

The structure of this study reflects the desire to analyze defense policy-making within a well-defined frame of reference. In the second chapter the evolution of United States' nuclear strategy is examined. For the sake of convenience three phases are distinguished - 1945-1954: The pre-massive Retaliation Era; 1954-1961: The period of New Look; 1961-1968: The Kennedy-McNamara Reappraisal. The purpose of this chapter is to provide the necessary background for understanding the nature and the changes of American nuclear strategy, and the impact of these changes on the United States' present defense posture. Useful insights into James R. Schlesinger's "doctrine" can be derived from this.

The third chapter considers the pressures and prescriptions for change in the early 1970s in terms of the rationale as well as the impracticability of "Assured Destrution". It reviews the political setting of the 1970's Detente; targeting philosophy; the dissatisfaction left by SALT I; the incredibility and inmorality of nuclear deterrence wholly based on holding cities as hostages; and, finally, the incompatibility and inflexibility of this doctrine in pursuit of its national objectives. In short, it establishes the immediate background to Schlessinger's innovation. An effort has been made to identify the main underlying domestic pressures on defense policy formation. Special attention is also given to Schlessinger as a critic of defense before the assumed office.

The fourth chapter conveys the essence of Schlesinger's doctrine: "Selectivity and Symmetry", and seeks to evaluate the contribution of Schlesinger's doctrine and to assess it.

Finally, one should emphasize at the very outset that sources on Schlesinger and his thought are still fragmentary. As Lawrence Martin has put it, "here is an area in which one would like to see Mr. Schlesinger's thought further elucidated".[14] I have to rely heavily on various reports by the Secretary of Defense and material from his press conferences, supplemented by Presidential Reports to Congress on foreign affairs - the guideline of defense policy.

FOOTNOTES (Chapter 1)

1 John Baylis, "Defense Policy Analysis: And the Study of Changes in Post-War British Defence Policy", International Relations, November 1973, 385.

2 R.E. Osgood, NATO: The Entangling Alliance, Chicago: University of Chicago Press, 1962, 5.

3 H. Kissinger, Nuclear Weapons and Foreign Policy, New York: Harper and Row, 1957, 7.

4 Message to a Joint Session of the Congress, March 12, 1947. In Documents on American Foreign Relations, 1947. New York: Council on Foreign Relations, 1947, 646-650.

5 "The Sources of Soviet Conduct", Foreign Affairs, July 1947, 581.

6 U.S. Foreign Policy for the 1970s. A Report to the Congress by Richard Nixon, President of the United States, May 3, 1973, p. 8.

7 Richard Nixon's Report, p. 9.

8 See Bernard Brodie, "Strategy and National Interests: Reflections for the Future", 1971 by National Strategy Information Center Inc. in the chapter "Vital Interests: By Whom and How Determined?"

9 Samuel P. Huntington, Strategic Programs and the Political Process (The Common Defense), New York, Columbia University Press, 1961.

10 Arnold Wolfers: Discord and Collaboration, Baltimore, John Hopkins Press, 1962, 47.

11 J.C. Garnett, "The Analysis of Threats", International Relations, November 1971, 1003.

12 J.D. Snyder, "Threat-Perception and the Armament-Tension Dilemma", Conflict Resolution, Vol. ii, No. 1.

13 H. Kissinger, "The Problems of Limited War" in Robert J. Art and Kenneth N. Waltz (eds.) The Use of Force, 99. Little Brown Company, 1971. Library of Congress Catalog Card No. 76-141 911.

14 Lawrence Martin, "Changes in American Strategic Doctrine--an Interpretation", Survival, July/August, 1974, 164.

CHAPTER 2

AMERICAN STRATEGIC DOCTRINE 1945-1970

To penetrate to deeper levels of understanding and meaning of United States' stragegic doctrine there is a clear need for an historical perspective. The transformation of nuclear monopoly to parity inflicted the painful burden on the United States to redesign her world role. This in turn necessitated the restructuring and redefinition of her strategic doctrines. However, no review of strategic doctrines can ever claim thorough coverage. My concern in this chapter is to put into various complementary perspectives the conflicts, consensus, consistencies and inconsistencies operative in the period from 1945.

The Pre-Massive Retaliation Era (1945-1954): In search of a nuclear strategic doctrine

Post-War Europe

Determined not to repeat the errors of its isolationist past, the United States committed itself to the formation of the United Nations as early as 1943. In April 1945 Russia's penetration of Europe upset Western calculations. The priorities of Allied wartime policy and Germany's total destruction had permitted the Soviet Army to challenge the traditional European balance of power. Military dominance enabled the Soviet Union to impose its will on the States of Eastern Europe. What is more, Stalin had threatened to interfere in post-war East European politics to ensure pro-Soviet governments along Russia's western periphery throughout the war years. Thus the Kremlin presented the United States and Britain with the choice of either recognizing Soviet political and strategic interests in Eastern Europe or accepting the post-war disintegration of the wartime Grand Alliance as the price of clinging to their principles of self-determination.

Nobody really knows how the cold war began. Twenty-five years of scholarship have produced no consensus. But the great confrontation between the United States and the U.S.S.R. since 1945 has been the central fact of international life. After the war American policymakers had to come to terms with the collapse of the European balance of power and the advent of nuclear weapons. 'These two developments have completely changed the strategic significance of the earlier components of American security: Allies, time, the overseas balance of power, oceans and industrial superiority.'[1]

The collapse of the European balance of power demanded a re-adjustment of the American role. The sense of national 'destiny' helped shape that adjustment from the start. In his message to a joint session of the Congress, on March 12, 1947, Truman provided the guideline for the future: 'Collapse of free institutions and loss of independence would be disastrous not only for [Americans] but for the world.'[2] The Vandenberg Resolution of June 11, 1948 is similarly clear evidence of the United States at a cross-roads.

> If the Marshall Plan may be likened to enlightened application of World War I loans, then military aid to the Western Union would be an improved version of World War II lend-lease, with the important distinction that weapons and supplies were intended to deter attack rather than to repel invasion after it had begun.[3]

In responding to these imperatives there seemed no alternative to an alliance in which the United States could join Western Europe and reinforce the purposes which both the Marshall Plan and the Brussels Treaty were formulated to serve. By a vote of 82 to 13 the Senate approved the North Atlantic Treaty on July 21, 1941. Collective Defence replaced the isolationist tradition. A new strategy--containment--was gradually devised. Whether containment represents anything new in American policy is a matter not only of observation but also of judgement. 'If one focused on one's own territory, this was labeled self-defense; if one focused on likely threats to one's territory, it could be labeled containment.'[4] The post-war situation was clear to the United States: that her shpere of finance was under Soviet threat. In essence, then, she rejected a strategy of no prior commitment to a strategy of prior commitment, which hopefully, would prevent a European (or other) war in the first place.[5]

The policy of containment was backed up by her monopoly of nuclear weapons in 1945-1949. Until 1950, United States' strategy envisaged fighting a nuclear war as well as deterring it. But there was much discussion of the merits of developing a strategy of total nuclear deterrence, and rejecting military planning which depended to a large extent upon conventional warfare concepts as developed in World War I and World War II.

This was all well and good but the strategy did not incorporate any appreciation of the fact that, once the Soviet Union had achieved the capability to attack the homeland with nuclear weapons, American security would be in danger of erosion through loss of nuclear monopoly.

Such monopoly ceased after the explosion of a Soviet nuclear weapon in August, 1949. Meanwhile most professional strategists were of the opinion that the use of the nuclear weapon alone could not defeat a nation which deployed vastly superior ground forces. Thus, the signing of the NATO treaty was a recognition of the fact that the West European powers alone could not match the Soviet Union's conventional strength and of the desire of these powers to link American nuclear forces to the defence of Europe. The concept of a relationship between conventional and strategic nuclear warfare was produced.

Early notions of nuclear retaliation as a way to counter superior Soviet conventional armed forces rested on the assumption of an atomic monopoly. The detection of a Soviet nuclear explosion in 1949 threw into doubt reliance on deterrence by nuclear retaliation. A requirement for a conventional build-up was predicted in NSC-68 an early 1950 policy paper (which did not, however, assess the acceptability of the economic costs involved).[6]

Although the result of a comprehensive survey of American strategic policy by the National Security Council, NSC-68, was relatively brief, the document examined and compared Soviet and Western capabilities, ranging from the conventional forces to nuclear weapons and included a projection of future nuclear capabilities and economic strength that caused much disquiet. 'The conclusion of the document was a real call to action, claiming that more, much more, needs to be committed in effort and resources by the United States if the American strategic situation were not to deteriorate steadily.'[7]

The document focused mainly upon the analysis and definition of national security goals, and is noteworthy for its systematic and comprehensive analysis of national strategy. It considered various alternatives with the weight of rational argument. But it was not overloaded with estimates of the force levels and weapons and budget totals needed to carry it out. But the fiscal 1951 budget makers of defence had to produce details about means, not least to satisfy congressional questioning. Against this background, any criticism of the defense budget's inadequacies based on this rationality of NSC-68 would not be well found.[8]

Since the drafters of NSC-68 concentrated on general considerations of strategy without facing budgetary restraints and the problem of reconciling the desirables with the possible, the whole report was held at the time irrelevant--unless unforeseen events occurred. However, the outbreak of the Korean War validated NSC-68 and its influence came to bear on the later development of American strategic doctrines.

The American military budget now rose to an annual level of $50 billion (by 1953), more than three times what it had been prior to the Korean War. Part of the increase reflected the costs of that War, which was prolonged far beyond 1950. Most of the extra appropriations, however, went to accelerating the expansion of the Strategic Air Command (SAC)--the primary repository of American nuclear capabilities. Other budget accretions went to conventional forces outside the Korean theater, including four additional American divisions deployed in Germany in 1951.[9]

In short, the 'pre-massive retaliation' era foreshadowed later developments in future United States' Strategic doctrine. The inadequacy of Western European conventional forces was safeguarded against the Soviet conventional capability by the American nuclear guarantee in her years of atomic monopoly. With the Soviet admittance to the Nuclear Club, the emphasis on conventional forces was more pronounced. The NSC-68 was already on the table.

The New Look (1954-1961: 'Massive Retaliation' --a viable Nuclear Strategy?

The New Look Doctrine stemmed from the recognition that nuclear weapons created, in Secretary Wilson's words, 'New relationships between men and material which emphasize airpower and permit economies in the use of manpower'.[10] The doctrine of massive retaliation, which threatened the use of nuclear weapons for a wide range of contingencies, thus had its roots in the desire for a balanced budget. Glenn Snyder's case study of the FY 1955 budget convincingly demonstrates that the decision to rely heavily on strategic capability emerged from the debates over this budget, which resulted in the reduction in the cost of preparing for war.[11] In terms of capabilities, the New Look meant a greater relative reliance on strategic nuclear air power and a 'de-emphasis' of land, naval, and tactical airforces.

The theory behind the New Look was developed in a study by the National Security Council, and the principle was supported by Secretary of State, John Foster Dulles who, in a speech in January 1954, announced that in order to get a 'maximum deterrent at a bearable cost' the government had desired to:

> . . . depend primarily upon a great capacity to retaliate, instantly, by means and at places of our choosing . . . We need allies and collective security, but our purpose is to make these relations more effective, less costly. This can be done by placing more reliance on deterrent power and less dependence on local defensive power.[12]
>
> [emphasis added]

But the speech by the Secretary of State at once invited public outcry. The memorable phrase became so deeply embedded in the public mind that many considered the Secretary a man who enjoyed brandishing nuclear weapons and who had lost his sense of selectivity and flexibility. Dulles' gloss on this statement of policy--namely that 'local defenses must be reinforced by the further deterrent of massive retaliatory power'[13] commanded less attention.

To question whether the strategy of 'massive retaliation' makes sense is to debate whether the United States actually possessed adequate diversification of forces at that time and whether it had, in fact if not in word, locked itself into options of nuclear holocaust or inaction. Was the strategic doctrine of massive retaliation feasible?

In 1949, when asked to what extent the atomic bomb deterred the Soviet Union, Dulles had responded: 'I don't think it is a decisive factor--although that's only a guess'.[14] But he was a diligent student of the concept of deterrence and became convinced that any open Soviet aggression would probably be a miscalculation and that official actions and utterances should leave no room for such misclaculation in the minds of the Soviet leaders.

The basis for his 1954 speech, and thus the rationale of the massive retaliation doctrine, derived from this study in deterrence and the experience of the Korean War 1951-1952. These principles, summed up in the following propositions, are brought out in Dulles' statements or articles over the years.[15]

> The strategy and power of deterrence had to be openly organized and more clearly defined to avoid miscalculation; and deterrence as a political-military-psychological concept must be developed.
>
> Effective deterrence depended upon allied co-operation. American troops in Europe should be recognized as a part of a deterrent power.
>
> The United States should organize its deterrent power in a manner which would assure allies and political adversaries that the United States would never take the military offensive.
>
> Massive retaliation power related only to aggression which itself constituted or would inevitable lead to general war.
>
> Local defenses and conventional forces should not be abandoned, for limited

> conflicts would require limited responses.[16]

In several respects Dulles was merely reiterating the policy of the Truman administration, which had assumed that deterrence against any major attack (either by Soviet Union or China) should ultimately be based upon nuclear weapons, air, and sea power. At the same time, he believed that extended deterrence could be accomplished by establishing a will and determination not to ignore open aggression. On the other hand, Dulles differed in approach. He argued that the potential adversary had not been 'given the kind of warning which is necessary for a deterrent to be fully effective'. He thus concluded:

> There is an urgent need to clear up the present vagueness by a free world decision which the potential aggressors cannot misjudge, either as to its power or its certainty--and 'certainty' means not only a statement of the circumstances of aggression under which punishing power would surely be invoked, but also certainty that it would never be invoked except under such circumstances.[17]

Dulles was perhaps the first 'cold war warrior' and the first Secretary of State to think in terms of overall strategy and posture and their implications for foreign policy. To Dulles the 'Massive Retaliation' formulation meant that the relatively unstated and underdeveloped strategy of deterrence of the previous administration should now be declared and defined.

Contrary to some interpretations, the espousal of 'Massive Retaliation' did not mean that the United States intended to rely _totally_ on strategic power and would reply to virtually any form of aggression with nuclear weapons. A careful reading of the COuncil Speech itself shows the error of such interpretations and criticisms; and that Dulles was never wholeheartedly committed to an extreme doctrine. Thus, he wrote,

> Local Defense will always be important. But there is no local defense which alone will contain the mighty power of the Communist world. Local Defenses must be reinforced by the further deterrent of massive retaliatory power. A potential aggressor must know that he cannot always prescribe battle conditions that suit him.[18]

Adlai Stevenson criticized the policy as presenting the United States with 'the grim choice of inaction or a thermonuclear holocaust'. It therefore invited Moscow to 'nibble us to death'. Dean Acheson asserted that such a strategic doctrine could only invite Korean-type war.[19] More serious criticism came from the academic world. Henry Kissinger found that 'Massive Retaliation' contained a tendency towards a 'Maginot mentality' or 'Fortress American' posture. He concluded that 'all or nothing' military policy 'makes for a paralysis of diplomacy'.[20] William Kaufmann's Military Policy and National Security, the classic critique of Dulles' 'Massive Retaliation', discussed a better, more credible deterrent in a new parity situation. James King, Robert Osgood, Wohlstetter, Bernard Brodie and others joined in the chorus.

To recapitulate the above argument, the New Look was a doctrine, i.e. a set of ideas and concepts concerning the mobilization, deployment and use of force as an instrument of deterrence and defence. But it was also a set of actual and planned changes in military establishment to strike a balance between the competing goals of national security and economic welfare.

Given the increasingly stringent budgetary limitations and the administration's strategic orientation all the Services stressed nuclear strategic functions at the expense of conventional capabilities for fear that concentration on the other would put their budgets at risk. The virulence of the rivalry for strategic missions--particularly for weapons associated with ballistic missiles, continental defence and missile defences--resulted in the wasteful duplication of development effort. The inability of the Secretary of Defense to choke this off is amply demonstrated in the air defence case by both the Nike-Talos conflict and the Nike-Bomare controversy. Meanwhile, the Services' desire for rapid requisition of new armaments--both in competition with other services (the first weapon to be operational would 'win the mission') and in the Cold War competition with the Soviet Union, resulted in the adoption of a 'weapon system' or 'systems' approach to military research and development (of which more later).

Nonetheless, factors beyond the control of the Secretary or the President (e.g. the interorganizational rivalry among the Services; the Soviet test of their own hydrogen bomb in 1953) significantly affected timing, tactics and even the extent to which the policy was understood. 'Massive Retaliation' is too often misinterpreted. In fact, the recipe for 'sound policy' uttered by Dulles himself could summarize the analytical approach to the policy of New Look,

> Sound policy cannot be derived from a priori concepts, whether moralistic, materialistic or legalistic . . . Our national interest requires that we relate our policies to the actual conditions of the world about us. This means that we must understand both the realities we confront, and the limitations on our power.[21]

The Kennedy-McNamara Reappraisal 1961-1968

Knowing where to begin is the most difficult step in any effort to develop a generalized statement of any policy formation process. To understand the McNamara Doctrine, it is necessary to understand the guideline given to him by President Kennedy:

> When I took office in January 1961, President Kennedy instructed me to: (1) Develop the force structure necessary to meet our military requirements without regard to arbitrary budget ceilings, (2) Procure and operate this force at the lowest possible cost.[22]

The instruction was accepted with relish, predetermined arbitrary budget ceilings were explicitly rejected. 'Today we examine all our military needs, and then decide at what point our military strength is in balance with the requirements of our foreign policy.'[23]

In discussing the 1950s defence policy, two factors--stringent budget and 'strategic monism'--combined with service autonomy and the use of budgetary ceilings, produced 'virulent service rivalry' in which services stressed the strategic mission to the exclusion of other functions. The Kennedy administration's desire for coherent military strategy logically implied the imposition of central control in order to monitor the activities of the services. To examine military needs and decide at what point military strength is in balance with the requirements of foreign policy is to accentuate the need for Central Control.

In the 1960s, 'strategic monism' was replaced by 'strategic pluralism', with the emphasis on creating 'multiple options'. This ended the reliance upon strategic capability for all contingencies in order to avoid a 'choice between humiliation and holocaust' in crises.[24]

In its quest for alternatives to 'humiliation or holocaust', the Kennedy administration fostered the strategic Concepts of Counterforce, Damage Limitation, and later Mutual Assured Destruction. The successive elaborations of the 'McNamara Doctrine' viewed separately under their fashionable labels, could not be described as major departures from past policies. Together, however, the changes promised to lessen the destructiveness of a general nuclear war in case deterrence failed. Actually, the United States, or any nuclear power, can never make a clear choice between strategies; it is, as a matter of fact, a question of emphasis. Since writers must write about and commentators must comment on nuclear strategic doctrines, they are usually given names--Dulles' Massive Retaliation; McNamara's Doctrine of Flexible Response; and the Schlesinger Doctrine of Selectivity and Symmetry. These complex policy formulations are not as simple as their lables suggest. The prescription for a viable strategic doctrine given by General Eisenhower in 1952 to Dulles could well be recalled at this point: 'Exclusive reliance upon a mere power of retaliation was not a complete answer to the broad Soviet threat.'[25] Dulles in the article "A Policy of Boldness" which appeared in Life also helps to classify the confusing apparently conflicting ideas embodied in various American strategic doctrines.

> The United States . . . should maintain a strong, balanced force--including land, sea and air power--capabile of expansion in any direction that seems wise. We should not put all of our eggs in one basket. But the abstract principle of balance should never be applied so vigorously as to impair emphasis on what is the practical need of particular circumstances. . .[26]

The McNamara Doctrine was a response to the contemporary international scene. The Missile Gap (1957-1961)[27] turns out to have been the other way round. The American strategic arsenal of thermonuclear delivery systems had probably increased so much as to leave the Soviet Union a poor second. But that was not the contemporary view. Near parity was presumed. Hence, 'Massive Retaliation' was criticized as an incredible extended deterrent. Meanwhile, Western Europe, the Soviet Union, and China were all challenging the United States for goodwill in the Third World. The multiple challenges imposed an obligation (given the 'destiny') to have the proper military forces and grand strategy to meet the whole spectrum of international problems. The problems called for balanced forces, 'which will eventually lead to balanced spectrum, of mutual deterrence at all levels of the political-military challenge'; and dominant nuclear forces, 'which will then lead to extended deterrence'.[28]

The best statement of this doctrine was made by Alain Enthoven on 10 February 1963, in which may be discerned antecedents of the Schlesinger Doctrine of the 1970s:

> Our approach is based on options, deliberations, flexibility and control. Rather than decide ahead of time which targets must be hit by which weapons, and then commit ourselves to it, our approach is to give the President a range of choices, so that he can select the plan whose targets and timing of attacks are most appropriate to the circumstances at hand.

The 'new' stress upon the balanced forces and upon comprehensive capabilities meant that the Air Force could no longer be permitted to neglect the tactical air power needed for close support of ground troops or the airlift needed for troop movement. Meanwhile, the Navy's stress on nuclear attack forces--Aircraft Carriers and Polaris Submarine--which had resulted in a neglect of anti-submarine warfare and escort forces, had to be corrected.

The Army too, which had so reduced the quality of its divisions that a war of more than a few weeks might have exceeded their capacity, had to re-examine priorities under the new administration. So defence spending rose. Between 1960 and 1965 the budget increased by $7 billion, with the following return in military output:

> A 45% increase in the number of combat ready Army divisions, a 45% increase in combat helicopters, a 100% increase in airlift capacity, a 57% increase in the number of Air Force fighter squadrons, a 100% increase in naval ship construction, a 100% increase in the size of counter-insurgency forces.[29]

Central to the doctrine, in other words, was the motion that the larger the forces, the greater the flexibility. Sufficient flexibility permits a choice of strategies. However, a further element in the impetus to provide such a spectrum on the conventional level arose from a felt need to revalidate 'extended deterrence'.

'Extended deterrence' in its 'purest' form was first propounded by John Foster Dulles in his 'Massive Retaliation' speech. As a strategic concept it was elaborated at the RAND Corporation and popularized by Henry Rowen in his 1960 Study Paper Number 18: National Security and the American Economy in the 1960s for the Joint Committee on Economic Development. Although it is difficult to establish the precise relationship between the McNamara Doctrine and Henry Rowen's paper, Herman Kahn's On Thermonuclear War and Richard Fryklund's The Hundred Million Lives, one could argue that on the basis of American nuclear superiority, the academic strategists had broadened the function of American strategic nuclear forces to deterrence of attack on the United States or a massive Soviet conventional aggression in Western Europe, to cover most other forms of aggression or provocation.

Having done so, an inescapable burden of world leadership, reinforced by 'destiny'[30] posed a dilemma: despite all rational efforts to prevent a conflict deterrence might still fail. Thus it was important to have a general nuclear war strategy as the last resort. And yet in language that has often been quoted: the administration felt impelled to get 'the nuclear genie back in the bottle' by broadening United States' capabilities. (This had been clear since as early as Kennedy's first State of the Union message in January 1961). The strategic Concepts of Counterforce, Damage Limitation and Assured Destruction, which with varying emphasis were at the heart of the McNamara Doctrine, were designed to close the gap between declared policy and likely action policy. Whether the Doctrine constitutes a viable nuclear posture, no one can ever know. The controversy over, and various interpretations of, the McNamara Doctrine, replete with statements stressing the pressing need for a viable defence or emphasizing the impossibility of achieving an effective defence or providing dramatic distinctions between First Strike/Second Strike Capabilities with First Strike Doctrine--could not be taken seriously. As Michael Brower put it, 'In nuclear strategy what we do is vastly more important, in the final analysis, than what we say.'[31]

In the official document "This is Counterforce" (7 February, 1963), it is clearly stated that

> In truth, it is a very old strategy, perhaps as old as the history of war, and its application is absolutely fundamental to any level of military conflict. Most people will recognize that pitting one's military force against the military forces of the opponent is the usual way of conducting limited warfare.

The Counterforce strategy, which many interpreted as a No-cities strategy, gives rise to great confusion. There were indeed 'two component parts' of the strategy that McNamara propounded: 'Counterforce' and 'Cities' (or No-cities).

In 'Counterforce' the principle is to go to the enemy's military forces while in No-cities, the principle is to leave his cities alone.[32] In nuclear war, targetting cities is a 'reprisal' or 'coercive' measure to force an enemy not to adopt a city-destroying strategy of its own. In the words of McNamara 'to give a possible opponent the strongest imaginable incentive to refrain from striking U.S. cities'. A strategy of striking cities with nuclear weapons constitutes a desperate attempt to achieve deterrence through the clear ability to gain total victory. But by its very nature, a counterforce strategy is a flexible strategy for fighting war --from small-scale counter-guerilla engagements to the highest global destruction. To equate counterforce with 'no-cities' is to confuse the overlapping different strategic concepts. 'The two notions complement each other, of course, in that both are intended to keep the enemy from using his weapons against us, one through forcible disarmament and the other through continued deterrence.'[33]

To keep the enemy from using his nuclear weapons against United States' cities and to extend deterrence over Allies (especially in Western Europe) could only be achieved by nuclear superiority. On October 21, 1961, Deputy Secretary of Defense Rosewell Gilpatric summed up the ratio of American to Soviet strategic nuclear forces:

> The total number of our nuclear delivery vehicles, tactical as well as strategic, is in the tens of thousands; and of course, we have more than one warhead for each vehicle. Our forces are so

> deployed that a sneak attack could not effectively disarm us . . . In short, we have a second-strike capability which is at least as extensive as what the Soviet can deliver by striking first.[34]

On May 15, 1962, the Wall Street Journal reported that 'The missiles now ready to retaliate against nuclear attack include 54 relatively vulnerable above-ground Atlases, 18 underground Titans and 128 elusive submarine carried Polaris weapons. In contrast, the bomber force boasts 630 heavy B-52s, 900 aging medium B-47s and 90 supersonic B-58s--half of which are kept to scramble after 15 minutes notice'. Deputy Secretary Gilpatric summed up the U.S.-Soviet position by 1965:

> As a result of our strategic Force Study, we now have in our planning, at least as far as 1965, a pretty definite force structure. We will have nearly 950 bombers (apparently all but about 200 B-47s will have been phased out) carrying, we hope, 800 air-to-surface missiles including Skybolts as well as Hound Dogs. We will have 1500 ICBMs operational, including Atlases, Titans, Minutemen and Polaris. We will have more than double the number of alert weapons that we have today by the end of 1965 . . . In other words, we will have twice the striking power by 1965 that we have at the end of the fiscal year (June 30, 1962). This is why we feel that no matter what the Soviets can do, based on the intelligence we have today, that we will maintain the margin of superiority that we possess today.
>
> [May 2, 1962, News Release by Deputy Gilpatric]

Thus, according to some[35] denials or no, the posture outlined in the Ann Arbor Speech delivered by McNamara, appeared to confirm the view that henceforth balanced forces, although useful, were of secondary importance. 'To sophisticated and unsophisticated alike, it seems that the United States, for the first time in her history, has officially adopted an aggressive first strike capability.'[36]

However, the above argument confused a first-strike doctrine with a doctrine for the first use of nuclear weapons, a counterforce capability with a first-strike capability and both with the utility of strategic nuclear superiority.[37]

> A strategic counterforce capability is a capability to use strategic weapons against military targets, either tactical or strategic. The requirements for a first-strike capability includes those for a counterforce capability plus additional requirements, namely: the intelligence or continuous surveillance capabilities to acquire a large number of targets at the same time and the firing discipline to launch a substantial fraction of one's weapons in a single co-ordinated attack or, in the case of missiles, in a more or less simultaneous salvo.[38]

In fact, neither a counterforce capability nor strategic nuclear superiority need be destabilizing if there is no advantage in a simultaneous or salvo attack and hence no incentive to be the first to launch one. Moreover, a first-strike capability may be inherent in the second-strike capability.

Anyhow, counterforce was regarded as a valid military strategy so long as it could stand any technological breakthroughs. It is based on the relative invulnerability of hardened or concealed missiles and the terror threat that, though cities are spared, they are yet held hostages somewhere along the line of the options. The counterforce strategy is 'dynamic' in its implementation, emphasising as it does controlled responses, with the hope of initiative in any conflict lying with the United States.

The technological breakthrough did appear. Early in 1963, McNamara stated that the Soviet nuclear strategic forces were becoming increasingly invulnerable even to a very large attack. He concluded that 'We are approaching an era when it will become increasingly improbably that either side could destroy a sufficiently large portion of the other strategic nuclear force, either by surprise or otherwise, to preclude a devastating retaliatory blow. This may result in mutual deterrence.'

This statement implied that the United States must give up the reliance on strategic nuclear weapons as instruments of policy. Bladwin[39] interpreted the invulnerability of missiles this way, 'Moscow, like Washington, is trying to make its nuclear deterrent and retaliatory power less and less vulnerable to surprise attack. As the invulnerability of missile launching sites increases and each side finds it impossible to knock out the others nuclear capability by a surprise first strike, the stability of the deterrent is expected to increase.' But such deterrence would have left credibility of extended deterrence in doubt as it remained until further elucidation by Schlesinger.

Damage Limitation and Assured Destruction are two sides of the same coin; they bring the problem of general nuclear war within 'a single analytical framework'.[40] Damage Limitation is neither, as Mr. George E. Lowe[41] argued 'the antithesis of a balanced force structure, nor associated with a first strike capability', nor 'synonymous with active deterrence'; it was simply 'the capability to reduce the weight of enemy attack by both offensive and defensive measures and to provide a degree of protection for the population against the effects of nuclear detonation.'[42]

Indeed, there had been discussions of this two-sided coin concept before 1965. But it was in February 1965 that the McNamara Doctrine spelled out the meaning of Assured Destruction: that is the vital and paramount capability--clear and unmistakable--to 'destroy an aggressor as a viable society even after a well-planned and executed surprise attack on U.S. strategic forces.' The Soviet Damage-Limiting problem was the American Assured Destruction problem and vice-versa. Thus, Damage Limitation and Assured Destruction do not specify to a particular doctrine, strategy or tactic. In point of fact, they are the objectives of a general nuclear posture in which the aim is to deter a deliberate nuclear attack upon the United States and its Allies by maintaining a clear and convincing capability to inflict unacceptable damage on a first-strike attacker, and to limit the damage inflicted upon the population. Thus, 'Damage Limitation' is the second mission of the strategic forces in case of a general nuclear war; and 'Assured Destruction' is to deter the Soviet Union as the operational process. Yet, could such a strategy, with the emphasis on 'Damage Limitation' or 'Assured Destruction' be viable? The situation stood in 1968 significantly unchanged from the way Secretary Gates phrased it, 'in order to maintain a valid deterrent we have to maintain a valid deterrent force capable of knocking at his [the enemy's] military power and not first bombing his cities. What we would actually do depends on circumstances, but we are adjusting our power to counterforce theory; or a mixture of counterforce theory . . . what we actually do in the period in a real operation is something else, I do not know for sure.'[43]

FOOTNOTES (Chapter 2)

1 See Introduction. Warner R. Schelling: Strategy, Politics and Defense Budget, Columbia University Press, 1962.

2 Message to a Joint Session of the Congress, March 12, 1947. In Documents on American Foreign Relations, 1947, New York: Council on Foreign Relations, 1947, 646-650.

3 Lawrence A. Kaplan, "The U.S.A. and the Origin of NATO", Review Politics, 31, 1969, 218.

4 George H. Quester, Nuclear Diplomacy, the First 25 Years, Dunellen, 1970, 24.

5 See William T.R. Fox, "American Foreign Policy and the Western European Rhineland". Proceedings of the American Academy of Political Science, January, 1948, 74.

6 Neville Brown, Nuclear War, London: Pall Mall Press, 1964, 70.

7 Schelling, Hammonds and Snyder, Strategy, Politics and Defense Budgets, Columbia University Press, 1962, 70.

8 See Schelling, Hammonds and Snyder, Chapter 2. For a comprehensive account of the analysis of NSC-68.

9 See Robert Osgood, NATO: The Entangling Alliance, 77-78.

10 Snyder, New Look, 469.

11 Ibid., 469.

12 U.S. Department of State Bulletin, vol. xxx, No. 76, 25 January 1954.

13 Dulles' Address before the Council on Foreign Relations, "The Evolution of Foreign Policy",

January 12, 1954, in DSB, xxx, 108.

14 Dulles' Interview in U.S. News and World Report, XXVII, July 8, 1949, 30.

15 "North Atlantic Pact", Vital Speeches of the Day, XV, August 1, 1949, 617-24; "The Challenge of Today", DSB, XXIV, June 11, 1951, 935-37; "Security in the Pacific", Foreign Affairs, XXX, January, 1952, 175-87; "A Policy of Boldness", Life, XXXII, May, 1952, 146-60; "Far Eastern Problems", Vital Speeches of the Day, XVIII, June 1, 1952, 493-95.

16 Guhin, John Foster Dulles: A Statement of His Time, Chapter 10, 1972, Columbia University Press.

17 Dulles' Draft of "A Policy of Boldness" in Papers, March 8, 1952, 29-30.

18 Dulles in DSB, XXX, 108.

19 Quoted in Coral Bell "Atoms and Strategy", in F.C. Besham (ed.) Survey of International Affairs 1954, London: Oxford University Press 1957, 99.

20 H. Kissinger, "Military Policy: Defense of the 'Grey Areas'", Foreign Affairs, XXXIII, April, 1955.

21 Dulles' Draft of "Challenge and Response in U.S. Policy", in Papers, August 9, 1957, 3.

22 John P. Crecine, "Defense Budgeting: Constraints and Organizational Adaptation", Inst. of Public Policy Studies, Discussion Paper No. 6, Ann Arbor: University of Michigan, 1969, 17.

23 Samuel A. Tucker (ed.), A Modern Design for Defense Decision: A McNamara-Hitch-Enthoven Anthology, Washington, D.C. Industrial College of the Armed Forces, 1966, 24.

24 William K. Kaufmann, The McNamara Strategy, New York: Harper and Row, 1964, 87.

25 Letter from Eisenhower to Dulles in Papers, June 23, 1952, 3-5.

26 Memorandum of Conversation with Hanson Baldwin on "A Policy of Boldness" Article in Life, in Papers, June 23, 1952, 3-5.

27 See George Quester, Nuclear Diplomacy, Chapter 4, "The Missile Gap", 145-97.

28 U.S. Naval Inst. Proceedings, June 1963.

29 Milwaukee Journal, 29 June, 1969, 8.

30 See Part I of this Chapter.

31 Michael Brower, "Nuclear Strategy of the Kennedy Administration", Bulletin of Atomic Scientist, October, 1962, 20.

32 T.C. Schelling, "Controlled Response and Strategic Warfare, IISS, Adelphi Papers, 19, June, 1965, 5.

33 Ibid., 6.

34 Michael Brower. "Nuclear Strategy of the Kennedy Administration", Bulletin of Atomic Scientist, October 1962.

35 U.S. Naval Inst. Proceedings, June 1963.

36 Baldwin, "Neither Humiliation nor Holocaust", U.S. Naval Inst. Proceedings June 1963.

37 For detailed argument see Draft by Thornton Read "Counterforce Strategic and Nuclear Weapons', August 6, 1962.

38 Ibid., 8.

39 Baldwin, "Neither Humiliation nor Holocaust", 63.

40 Statement of Secretary of Defense Robert S. McNamara before the Armed Services Committee on the Fiscal Year, 1966-1970 Defense Program and the 1966 Defense Budget, February 18, 1965, 37-38.

41 U.S. Naval Proceedings Inst., June 1965, "Damage Limitation: A New Strategic Panacea?" 40-47.

42 Ibid. Testimony before the House Military Affairs Committee on February 18, 1965.

43 Hearings before Defense Appropriations Subcommittee, Department of Defense Appropriation for 1961, 86th Congress, 2nd Session, 1960, Part I, 26.

CHAPTER 3

PRESSURES AND PRESCRIPTIONS FOR CHANGE IN EARLY 1970s

No rational man would consciously advocate making 'Assured Destruction' the core of an international political posture. It is natural that in the second half of the 1960s critics of defense policies complained most vocally about the United States' strategic 'inmobilism'. Political considerations, therefore, as well as a genuine desire to provide a better strategic doctrine led to demands for a reconsideration of the Mutual Assured Destruction stance.

American political objectives of the 1970s, could be expressed--in formal declaratory terms--as 'building a peace'. The three foreign policy reports to the Congress[1] submitted by President Nixon assert that 'the post-war period in international relations has ended'[2] and that it is time to enter an era of negotiations. Thus a central aspiration of American policy became the translation of the Soviet-American nuclear competition into a 'less dangerous form of competition'.

Yet limits were set to the transformation process. W. and A. Durant in <u>The Lessons of History</u> state that 'peace is an unstable equilibrium, which can be preserved only by acknowledged supremacy or equal power'--a view which finds many echoes in American policy pronouncements. Thus the (then) Secretary of Defense designated Elliot L. Richardson wrote to Senator Stuart Symington on 23 January, 1973, 'I believe that to maintain "clear sufficiency" now and in the future, we must preserve a position of superiority over potential opponents in certain areas, such as in technology, becasue of the inherent nature of competition with a closed society. In other words, I would view "clear sufficiency" as an end point, with "superiority" in certain areas a requirement to achieve that end point.'

Be that as it may the new approach to Superpower relations demanded a reappraisal of the strategic doctrine of Mutual Assured Destruction. Reappraisal of what the United States was doing internationally and why it was being done was entrusted to the National Security Council. The NSC produced numerous study memoranda and exhaustive reports on topics ranging from the relatively specific to comprehensive reviews of posture and global commitments.

This chapter seeks to reveal the pressures and prescriptions for change: to examine the political setting in the 1970s and so identify the 'tradeoff' or 'bargaining' relationships in the era of negotiations between the two superpowers; to analyse the shortcomings and incompatibility of Mutual Assured Destruction and the search for a more practicable nuclear strategy; and to focus on the problems and perspectives that arose from United States' attempts to 'translate strategic doctrine from power to policy.'

The Political Setting: Detente in the 1970s

The Imperatives of Co-operation

As the decade of the 1970s unfolded, East-West relations made progress in the context of the long-term trends of international politics. The interests of both sides were judged to be best served by a process of detente. Continuing tension became tinted by soft-line diplomacy. It seemed that a more stable and adaptable international framework could offer a more viable order to moderate the East-West conflicts of the years since World War II.

Detente applies to relationships. 'It is the relationship which is, or is to be made, less tense and more relaxed, and this relaxation should involve not only tolerance but also a candid recognition of realities.'[3] The roots of this detente are to be found long before the 1970s however. In June 1963 President Kennedy had put out a feeler for 'relaxation of tensions' based on the common 'abhorrence of war'. According to Arthur M. Schlesinger, Jr., one of Kennedy's motives was to get a Soviet-American front against communist China which the President considered to be 'the long-time danger to the peace'.[4] Whether that motive was overriding or not there is no doubt that in Kennedy's entourage, detente was very much in the air,[5] and from these beginnings evoked the more ambitious detente of 1972.

A basic reason for the ambitious detente of 1972 was security policy. The superpower deternte has captured the mood of the United States' low-profile approach to international politics. Likewise, it fostered the hope of the recognition of the status quo by the Russians. In addition, Willy Brandt's Ost-politik was part of the German answer to the superpower dialogue and to the European inability to define common objectives of Western European security. These two dimensional developments, no doubt, eroded the consensus between the two superpowers with their respective allies and yet increased support from all parties concerned for the whole on-going process of Detente. The results are multi-faceted: Mutual Balanced Force Reductions; The Conference on Security and Co-operation Europe; Strategic Arms Limitation Talks.

Meanwhile, the technology of war has drastically changed the nature of military power. The Soviet Union has raced to rough nuclear equality with the United States. With this as background, it seems likely that 'parity' is always open to Soviet Union to use military force to advance political ends. Thus the rationale for Kennedy's detente was not very different from that adopted by President Nixon for his detente. By 1963, United States authorities believed that the United States and the Soviet Union were about evenly matched in strategic nuclear weapons.[6] Meanwhile the 1960s witnessed the rapid growth of Soviet military technology capabilities. While those of the United States grew, the American strategic build-up 'tapered off' by 1967, at the very time the Soviet Union accelerated. Kennedy in 1963 as much as Nixon in 1972 professed to be mainly concerned with slowing down the arms race. Even as early as October 1966 President Johnson already made a 'stab' at what he called, 'reconciliation with the East', but he never got far enough to talk of a more general detente.[7]

With the advent of the 1970s, both superpowers faced increased domestic restraints of a different kind. A changed temper of the American people began to make an impact on the Nixon administration. The Americans, as a result of the Vietnam conflict, came to question their 'destiny'--the premises, goals and priorities of foreign policy. The situation certainly appeared to demand self-consciousness and self-examination. Kissinger had described this mood as a flagging of the psychological drive, President Nixon spoke of 'growth among the American people of the conviction that the time had come for other nations to share a greater portion of the burden of world leadership.'.[8]

The unanticipatedly high costs in time and money for Soviet Union strategic build-up proved to be a strain on Soviet economy. It has failed to match up with the West in a fully well-developed modern industrial state. As early as February 1967, James Schlesinger already wrote that even the United States with an abundance of resources and a relatively high degree of flexibility in allocating resources often finds the projected capability is obtained at higher than predicted cost. 'The point is that other nations, lacking our abundance and our flexibility, are likely to be faced with the same question in a form more cruel.'[9]

In short, both the United States and the U.S.S.R. sought a detente in their relationships with each other in order to focus on other problems. In brief, 'detente is to create a vested interest in mutual restraint'.[10]

Before proceeding to the ongoing discussion, for the purpose of this essay--'Strategy is to translate power into policy'--it is beneficial for us to distil the realpolitik view of the world from Kissinger's published works and public statements (Kissinger, 1969, 1965, 1957). They seem to represent the less abstract thinking of the President and are confirmed by the direction and style of foreign policy under Nixon administration. In a nutshell, Kissinger appears to view political interplay among states in terms of power flows and contractions. Like Raymond Aron he holds that it is in the nature of states compulsively if not indiscriminately, to assert their influence and to secure their power position vis-a-vis other states. The task of foreign policy is to manage power. In fact, the terms of power in the world are better determinants of national interest than ideology but ideological aspirations could in some ways foster popular support. Thus, within this context, states are prepared to make use of military force as an instrument of coercion, intimidation, and a symbol of status while exhibiting great restraint in its actual employment. Therefore, early works reflected the idea that the United States should examine closely its own security requirements before undertaking any great power negotiations, and only enter into them with a guarded scepticism about results.

In this context the brutal reality of reconciling objectives in its foreign policy--strength, partnership and negotiations--Mutual Assured Destruction must have been far from convincing. It was incompatible, at least in declaratory posture, with the 'era of negotiation' by immorally holding the cities of the potential opponent as hostages. The 1965 posture statement (delivered in February) given by McNamara explicitly illustrated the strategic concept of Assured

Destruction and Damage Limitation:

> The strategic objectives of our general nuclear war forces are: to deter a deliberate nuclear attack upon the United States and its Allies by maintaining a clear and convincing capability to inflict unacceptable damage on an attacker, even were that attacker to strike first; in the event such a war should nevertheless occur, to limit damage to our populations and industrial capacities.
>
> The first of these capabilities (required to deter potential aggressors) we call "Assured Destruction", i.e., the capability to destroy the aggressor as a viable society, even after a well-planned and executed surprise attack on our forces. The second capability we call "Damage Limitation", i.e., the capability to reduce the weight of the enemy attack by both offensive and defensive measures and to provide a degree of protection for the population against the effects of nuclear detonations.[11]

The then existing strategic doctrine was incompatible with the 'era of negotiations' and did not provide flexibility to pursue the real-politik. No doubt, given the concept that foreign policy is to manage power, one would readily appreciate the message of President Nixon to the Congress in January, 1972, saying, 'No President should be left with only one strategic course of action, particularly that of ordering the mass destruction of enemy civilians and facilities. We must be able to respond at levels appropriate to the situation.'

Problem of Deterrence

Deterrence through Assured Destruction (MAD) was enunciated in the early 1960s and was adopted as an official doctrine in the 1965 Posture Statement by McNamara. Two controversial views were then expressed. While 'the defense intellectuals have generally defended Mutual Assured Destruction as the key to deterrence and demurred as the rethinking of basic assumptions and the adjustment to new perceptions of the role and utility of military force has proceeded on arms levels above those previously considered in the realm of strategy',[12] others condemned it either on moral grounds or on the credibility of such deterrence policy. 'It is a terrible strategy, and a very high priority, long-run objective should be to get rid of it altogether.'[13]

Perhaps the most fundamental objection is that the threat of massive nuclear retaliation in a limited conflict may not be credible. MAD may not convincingly deter an opponent from initiating nuclear warfare at lesser levels to achieve limited objectives. In fact, the most common argument sounds like this: 'Potential conflicts that involve important but not supreme national interests may induce one side to initiate the limited use of nuclear weapons, and pure MAD, whose failure must result in mutual national suicide, offers no credible deterrent to such an action.'[14]

Deterrence is basically an attempt by the deterrer to prevent the deterree from undertaking a course of action which the deterrer regards as undesirable, by threatening to inflict upon the deterree unacceptable costs in the event that action is taken. It is achieved by influencing the deterree's perceptions. As Glenn Snyder, in Deterrence and Defense put it, 'Deterrence works on the enemy's intentions.' Thus, any potential gains to be made must be outweighed by the costs which the deterree belives would be inflicted upon him in the event of taking the prohibited action.

Most important of all, the international order is to be jeopardized if once the concept of deterrence becomes ambiguous and the employment of it turns shaky. In short, 'deterrence is the most important and fundamental of the regulatory measures to control the use of force and to sustain international order in the contemporary era.'[15] This is not to say that states always act immorally in their exploitation of forces. Moreover, it does not follow that foreign policy is entirely a matter of selfish calculation. 'States, as a rule, are not exclusively concerned with the pursuit of specific benefits for themselves in an international setting dominated by the conflict of identically motivated competitors. States also concern themselves with the quality of the international environment itself.'[16]

However, a deterrence situation cannot exist without threats, tacit or explicit. And threats cannot exist unless it seems possible that the penalties concerned may be actually applied. Here comes the core of conflicting views between those who favored MAD and those who condemned it. 'How much deterrence is enough?' 'The greater the hurt, and the greater the perception of this hurt by the adversary, the nearer does the threatened penalty approach the absolute,'[17] is likewise logically repudiated by the argument 'to make punishment fit the crime'.

In fact, ingredients of deterrence expressed as capability, cost and intentions, identified by William Kaufmann[18] are still relevant today. To be deterred, 'a potential aggressor must be convinced that (1) the United States has the capability to do what it threatens, (2) carrying out the threat would inflict costs on the aggressor which would exceed the value of his objectives, and (3) the United States would in fact act as announced.'[19] Yet despite the prescription outlined in these criteria, deterrence is essentially a psychological phenomenon, not an exact science subject to mathematical analysis.

To paraphrase, Herbert F. York for further elucidation,[20] a strategy of nuclear deterrence is one to prevent certain political or military actions by threatening to use one's nuclear weapons rather than actually using them. 'Maintaining such a strategy, therefore, is more a matter of political psychology than of nuclear technology.'[21] A political adversary will likely be deterred if he believes that the nuclear punishment will be more severe than the achievement of some particular objectives. Therefore, the actual physical properties of weapons only enter the deterrence calculus in so far as the physical properties affect the beliefs of the various properties. And yet, 'if and when deterrence fails, the matter changes radically. Then it is no longer what people believe about the weapons that counts, but the real physical facts about their properties.'[22]

Regardless of the ambiguity with which deterrence can be ascertained there is little argument over what makes it work. If it works for twenty-five years or so, it does not mean that it will work again in the future nor does it guarantee that it would continue to work. Fred C. Ikle shook the academic world by his article "Can Nuclear Deterrence Last Out the Century?" which appeared in _Foreign Affairs_, January, 1973. But one has to admit that to question the wisdom of nuclear deterrence, either in the form of Mutual Assured Destruction or in the form of Controlled Reprisal is in fact to question 'the logic inherent in technology and the theoretical accomplishment of a whole generation of defense analysts.'[23]

However, one cannot ignore such questions even if they throw doubt on the wisdom of deterrence. Deterrence is the heart of nuclear strategy. Any strategic doctrine must fulfill the first prerequisite to deter the outbreak of a nuclear war. But deterrence is not a mathematical formula, it develops and broadens its parameter. Thus, any criticism on a strategic doctrine may in fact first begin with the deterrent criteria.

Deterrence: Finite or balanced of what?

'Assured Destruction' is to deter nuclear attack upon the United States through fear of reprisal. Reprisal would consist of atomic destruction of the economy of the attacking nation of sufficient magnitude to be considered as unacceptable by the government of that nation. This reprisal would be accomplished with the surviving United States' forces in sufficient quantities to inflict an unacceptable demographic and industrial destruction. Emphasizing 'Assured Destruction' through 'simplifying the equation of nuclear war'[24] and by directing the offensive forces ultimately against cities is often called basic or finite deterrence.

Assured Destruction required a military capability sufficient to ride out a massive surprise attack and still make the costs to the aggressor out of all proportion to whatever gains he may have in mind in initiating his attack. Such logic is taken to be highly expensive by the aggressor in deterrence value when one begins to talk about one's homeland as expressed by Brent Scowcroft: 'while credibility of intentions can never be so absolute as to go beyond question, the likelihood of response to an all-out attack on one's homeland is sufficiently high as to place doubt at an irreducible minimum.'[25]

As noted earlier, deterrence involves a necessary concern with the psychological outlook of the adversary or the potential adversary. Assured Destruction is a 'simple' strategy, simply explained: it eliminates uncertainty and it maximises penalities. Wolfgang K.H. Panofsky[26] rationalized the conversion of McNamara saying that 'during his tenure in office he gradually became convinced that such a policy was unworkable, both on physical and military grounds:

civilian casualties in connection with a purely anti-military attack were still apt to number in the many millions and one could not be sure the opponent would follow a city-avoiding strategy; instead he might choose to target centers of population.'[27] To paraphrase, one suspects that 'city-avoiding' strategy with its various options was held unworkable either on technological grounds or on the uncertainty on the part of the Russians. But options were already being discussed and explored. Accordingly, McNamara in his later years completely changed his 1962 position, moving toward a policy of deterrence on 'Assured Destruction'.

However, 'Assured Destruction' does not mean that the adversary cannot be deterred by being a 'simple strategy'. To use Colin S. Gray's words that 'we are unsure as to what actions would weaken deterrent effect, because we are far from certain as to which actions have deterrent effect,'[28] Mutual Assured Destruction, therefore, has a high deterrent value, certainly as high as is necessary for the massive pre-emptive strike in military capability for which it is operative.

Such claim may easily be verified by a careful review of congressional testimonies offered over the years by Herbert Scoville, Paul C. Warnke, George Rathjens, Jerome Kahan, Marshall Shulman, etc. Paul C. Warnke expressed the sound logic embedded in MAD in the Hearings of the first session of the 92nd Congress,

> I have never been able to arrive at any better concept than the concept of Assured Destruction as being the best deterrent of nuclear war, and I think it is because strategic nuclear weapons have to be regarded as basically unusable.[29]

Seen in terms of Kaufmann's elements of deterrence, the United States at that time still possessed an obvious capability (this will be further elucidated in the section on weaponry) to impose 'unacceptable' costs on the Soviet Union. Mutual Assured Destruction was in fact theoretically 'credible' as long as the Soviet Union could not be 100% certain that the United States would not do it. Additionally, threats of this sweeping kind have the attraction of economy for the deterrer, who may spare himself the expense of maintaining large conventional forces to be widely deployed in all quarters to meet possible infringements, or who may divert the resources urgently needed at the time (or conceived to be urgently needed) to a particular capability or to fight a particular local conventional war as in the case of Vietnam. One cannot have everything. Thus, to argue that Mutual Assured Destruction is incredible cannot be in fact proved or disproved because of the ever present element of 'uncertainty'. What one can really argue then is the inflexibility of Mutual Assured Destruction to 'translate power into policy', to 'provide moves in realpolitik'. Indeed deterrence is related to possible usage. We, therefore, must consider the 'uses' of nuclear weapons in evaluating their deterrent significance.

Dissatisfaction over the inflexibility is explicitly summed up in Colin S. Gray's phrase, 'Doctrinal Immobilisme'.[30] He expressed the dissatisfaction of the utility of such a strategic doctrine: 'Ingenious reasoning was the product of the inescapable paradox that western strategic theory--and policy--rested upon the promise of an act of destruction that no sane politician could ever reasonably invoke in execution.'[31] He continued to argue that 'strategists were not blind to the fundamental irrationality of the system of strategic relations to which they had contributed in so creative a fashion.'

His dissatisfaction on Mutual Assured Destruction would be more easily appreciated had he not been at the same time criticizing the academic strategists in general. However, the same dissatisfaction is superbly written by James R. Schlesinger while he was still a member of the staff of the RAND Corporation. What he wrote, in 1968, is of the greatest interest and significance in the light of his 1973 'innovation':

> Future forces should have sufficient flexibility built in so that they can perform something more than that strategic mission which appears dominant today. In the present context, with Assured Destruction as the sole criterion, we have been selecting hardware essentially for the purpose of the population kill. Yet, with the build-up of major devastation potential on both sides, it becomes increasingly doubtful whether such capabilities are the most reassuring or desirable to possess or whether they could serve any useful purpose in that range of contingencies which are becoming probable.[32]

Although he ruthlessly criticized the 'shortsightedness' of the then existing Mutual Assured Destruction doctrine, yet he realised the difficulty in policy-making. For this is not the only early innovation of the modes of thought that fostered the so-called Schlesinger doctrine. In December, 1965, he wrote,

> In the real world the tendency of governments to concentrate attention on immediate and pressing concerns makes it difficult to give adequate weight to insight. More allowance can be made in future, work for the alteration over time of opponents' objectives and strategies (particularly in response to our own moves) and for our own adaptation to these anticipatable, if not predictable, changes in behaviour.[33]

Again, in February, 1966, he wrote on the same theme,

> The gulf between the wars for which we prepare and the wrong wars that we fight

> at the wrong times and in the wrong places brings us back to the organizational image of the outside world and the nature of foes which, . . . tends to ossification and to obsolescence. Sudden changes in image frequently coincide with changes in administration, which in the United States are typically associated with "New Looks in defense" of one sort or another. While on the outside, the opposition broods on capabilities, possible wars, and the nature and behavior of foes, which the incumbents are presumed to be willfully overlooking. Not uncommonly the opposition is correct, but the views, nurtured on the outside become dominant on the arrival in power. When they become dogma, they may not only be out of date, but become increasingly more so, particularly when adhered to rigidly.[34]

At the same time similar views were being expressed by other critics with different tones of emphasis. Brent Scowcroft[35] stressed the development of Soviet retaliatory forces which had 'narrowed' considerably the nuclear utility enjoyed by the United States. The loss of the credible first-strike capability destroyed whatever utility massive retaliation may still have possessed and created doubt in some minds that the United States would be willing to come to the strategic defense of Europe. 'The key element of the components of deterrence appears at this stage of relative force capability to have become a particular aspect of intention: i.e. the objectives for which the U.S.A. would be willing to stake its survival.[36]

To put it more explicitly, although making preparations to fight at all levels less than total war and announcing these preparations to the Soviet Union would degrade the 'Assured Destruction' deterrent and thus tend to make a limited strategic conflict more likely, the alternatives that the United States enjoyed by such a doctrine are still pretty limited. It would place her in the position of having either to back down or to respond with an unrestrained attack. 'On balance, making preparations for the possibility of limited strategic conflict would seem to be more consonant with the principles of deterrence and to serve better the national interest.'[37]

General Bruce K. Holloway,[38] U.S.A.F., criticized the unconvincing aspect of the 'Assured Destruction'. 'Who believes we would do it? What sould it accomplish? Would this deter them?' He too prescribed that the United States 'want to deter all incursions against America--and against our allies of the Free World. If we have the flexibility of planning and application of nuclear weapons for meeting the pressures from the lowest levels of intensity against any type or system of targets, instantly or effectively, then the President will probably never have to fall up to such an awesome decision.'[39]

Raymond D. Gastil[40] shared the above view and labelled the 'Assured Destruction' as 'Doomsday Approach'. 'How can our leaders credibily imply that they would initiate nuclear war in a wide range of actions?' He then quoted the example of Quemoy. 'If our nation is to appear as though it will initiate nuclear war rather than lose even minor points such as Quemoy or Matsu . . . our nuclear threat, therefore, may well come down to being a threat of committing something close to suicide.'[41] Backing down in face of threat could enflame ambition and over confidence in opponents, and thus give birth to bitterness and irrational stubbornness in the leadership.

Donald G. Brennan[42] advocated the hypothesis that both the United States and the U.S.S.R. could continue to deter even as their 'Assured Destruction' capabilities shrank with the build-up of defensive systems. He emphasized (in an article written in June 1967) that a 'Soviet freeze' i.e. one on offensive systems and with no limit or deployment of defensive systems. His strongest argument is that 'both sides retain the ability to destroy say five to ten percent of the rivals' population; mutual deterrence will continue to be as effective as with 25% to 30% of the respective populations at risk, but we should all be better off--if deterrence were to fail'.[43] He crowned the basic fallacy embedded in the 'Assured Destruction' as that 'it tends to force us to a posture of "Assured Vulnerability".'[44]

Perhaps massive surprise attack on the United States is the least likely way in which nuclear weapons would first be used: but certainly not out of the question. Therefore, this must be the starting point to test the credibility of the strategy of Assured Destruction. An attempt to make policy 'usable' can only be viewed more fruitfully from that of 'intended' versus that of 'actual' effects. The starting point stressed by Fred Charles Ikle[45] is that we must consider not just today or tomorrow, but the indefinite future. He gauged methodically the problem of deterrence:

> An almost exclusive emphasis on deterrence could be defended as a satisfactory long-term policy if it could be convincingly argued that successful deterrence was tantamount to prevention of nuclear war. There exists no rational basis for such an argument. No matter how cataclysmic the threatened "Assured Destruction", those calculated decisions which our deterrent seeks to prevent are not the sole process that could lead to nuclear war. We simply cannot know which of the various potential causes is most probable--whether it be a coherently calculated decision to attack, or an "irrational" decision or technical accident.[46]

In contrast to Fred Charles Ikle who prescribed that to break the vicious circle in nuclear strategy is (i) to cast out the dogma that to deter nuclear attack, the threatened response must be the mass killing of people; (ii) to discard that response to nuclear attack must be the prompt, even instant launching of nearly the entire nuclear force and (iii) to eliminate the vulnerability of United States' strategic arms to surprise attack; Colin Gray seems to favour a more flexible use of tactical nuclear weapons after convincingly emphasizing the political nature of any nuclear strategy. To risk a platitude, he argued, it is governments, no people, that act in international politics. 'It is right (an embarrassing question)to threaten to execute/murder people by the tens of millions for the acts of their often not-elected representatives?'[47]

Herbert York[48] cast a gloomy shadow by saying that, 'if for any political psychological or technical reason deterrence should fail, the physical, biological and social consequences would be completely out of line with any reasonable view of the national objectives of the United States or the Soviet Union.'[49] He came to the conclusion of a general disarmament and arms control as alternatives to the Mutual Assured Destruction.

The same orientation is shared by Martin J. Bailey.[50] He argued that if deterrence of all-out war is unreliable, and if all-out war, however devastating, is likely to happen at least every hundred years or so, the death of the United States is assured within the foreseeable future. He criticized Mutual Assured Destruction and its underlying philosophy of trying to reduce the risk of all-out war through terror 'almost without regard to any other consideration'. Thus, he opted for greater defence through Damage Limitation as a prescription to the incompatibility of Mutual Assured Destruction by saying 'unless we protect our society to survive and recover'. No doubt, one finds him in the pro-ABM camp.

How much destruction would be considered 'unacceptable' to the Kremlin leaders is another favorite criticism on Mutual Assured Destruction. James R. Schlesinger made the following observation:

> These questions are so difficult, not only because Soviet intentions remain obscure and little light is necessarily shed on them by Soviet declaratory statements possibly issued with the intent to decieve, but because the Soviets themselves may not know how they should behave . . . "knowing your enemy" becomes a particularly challenging assignment, if the enemy does not know himself. One may not exactly require second sight, but the existing odd bits of political wisdom do not provide much illumination.[51]

Other critics also discussed this point. Brent Scowcroft questioned the level of damage an enemy will consider unacceptable--or, in terms used in the Assured Destruction definition, which will destroy his society as a viable entity is much more uncertain than any force capability calculation. He stated that the whole problem is 'highly abstract and subjective.'[52] Moreover, estimates of the amount of industrial damage and population loss the Soviet Union would consider unacceptable have not by any means been consistent over the years.

Donald G. Brennan, the most severe critic, quoted McNamara's statement to present the case on the irrationality and absurdity of the implication 'unacceptable'. He emphasized that McNamara recognized that what constituted 'an unacceptable degree of damage was not subject to precise specification, and in spelling out his requirement, McNamara said in his Posture Statement of 1968:

> In the case of the Soviet Union, I would judge that a capability on our part to destroy say one fifth to one quarter of her population (i.e. about 50 million Russians) and one half of her industrial capacity would serve as an effective deterrent.

The crucial question, however, as Colin Gray sees it, is how much retaliation is enough. Again, the unavoidable question follows: 'Enough for what?' 'Enough to deter, on some rough and ready putative deterrent calculus, or enough in quantity or quality to offer some prospect of thermonuclear war termination short of total catastrophe.'[53]

To simplify the above criticism, the ideal underlying such argument could be summed up as 'balanced deterrence'.[54] Balanced deterrence strategy is to emphasize damage limitation in view of the destructiveness of nuclear weapons and the reluctance of any nation to strike first. Its posture can be achieved by defenseive emphasis as represented by the foremost critic, Donald Brennan, in which superiority or parity is achieved through defensive rather than offensive systems. Most of those who favored an ABM system could be categorized under this heading. Others, like Fred Ikle prescribed a better control of nuclear weapons and challenged the academic strategists to think for a better alternative for the 21st century. Colin Gray, the realist, prescribed a more 'usable' tactical nuclear weapon so as to provide flexibility within the Mutual Assured Destruction. I am not sure where Colin Gray stands, it is apparent that both Brennan and Gray do not completely denounce the Mutual Assured Destruction. What they really advocated is to have greater flexibility together with a greater emphasis on defensive effort. This theme of thought is crystal clear in the case of James R. Schlesinger in which one recognizes the faint outlines of the United States' present strategic doctrine. What he actually criticized is lack of options beneath Mutual Assured Destruction. In other words, he does not renounce the whole strategy itself.

> My personal view is that the government has been more concerned to establish the firebreak than it should have been. That, however, is not my point, which is that the firebreak represents a constraint cutting across the multiple option approach.[55]

No better illustration of this subtle argument is available than Thomas C. Schelling's discussion on city-avoidance. He rationalized that nuclear war was not a shooting gallery where one had to shoot at something. The opponent's cities could be left for a while as hostages to his good behavior. 'The basic reason for not destroying cities is to keep them at our mercy'.[56] He reasoned that such a desirability did not depend on being actually able to disarm the opponent, since the intention was to bargain away his willingness to use capabilities he might still retain.

Given the competition for funds that always exist in the federal government, the 'priceless' quality of national security has led to an assymmetric competition for the marginal dollar between 'means' of national security (weapons systems, forces in being and so on) and the 'end-products' of, for example, welfare or education policy. Expressed more explicitly, security is 'priceless' in the sense that it is difficult to set a price on it. No cardinal measurement of security is possible; how much security is enough depends on subjective estimates of what makes one feel secure. 'Analysts say that the question is more of balancing more or less security with other goals; but there is no clear-cut way to relate higher or lower security levels with increments of the military means by which security is provided, much less weigh them against non-defense goals'.[56a] The vagueness of the defense end-product--the provision of security--and the concreteness of means--weapons systems etc.--has serious implications when considering the problem of allocating resources in the national budget.

Although we could not provide the solid proof that McNamara opted for 'Assured Destruction' was mainly due to the budgetary restraint, it might well be argued that the threat environment of the 1960s was considerably different: the Soviet strategic threat was no longer the only concern, thus making conventional forces important again and at the same time reducing the premium on a hasty development of strategic systems. In short, Arms Control became a salient issue in the 1960s and an acceptable political justification for delaying weapons' innovation.

Since one cannot have everything one wants, allocation of resources within 'defense budgeting' is always a difficult problem. The acceleration of Soviet ABM deployment became apparent in 1966. In November 1963 the Soviets had announced that they had designed an effective anti-missile; and a year later, the Galosh anti-missile missile was paraded in Moscow. McNamara's reaction, based implicitly on cost-effectiveness, was to procure offensive offsets (Poseidon and Minuteman III, with their multiple warheads) which actually increase 'greatly the overall effectiveness of our Assured Destruction force'.[57] He argued that since it was cheaper to procure offense than defense, the rational thing to do was to procure offense in reaction to an adversary's defensive deployment.

Since no one could establish a strong argument between the budgetary restraint and the decision for Mutual Assured Destruction, academic strategists and analysts seized on this point as part of their criticism on the strategy itself. 'Accommodated with budget-inspired rationalizing, they invented a new strategy--'Assured Destruction.' Perhaps this can be claimed as a virtue, but only in a sense comparable to the "argument" that the best automobile is one without wheels, because it is the safest and does not cause highway congestion."[58] However Johan J. Holst[59] rationalizes that the concept of 'Assured Destruction' may also have served as a useful political function at the time in terms of dampening incentives for stepping up the arms race by providing a barrier against 'internal American pressures' for greater offensive weapons arsenals. It provided a standard of nuclear sufficiency.'

The controversy over whether MAD is a result of budgetary restraint or a result of cost-effectiveness analysis or a prevention of the firebreak is summed up neatly by James R. Schlesinger who argued that the United States has refrained from making certain investments in the capabilities for low-level nuclear warfare, not only because of the expense but because the creation of such options might tempt the United States readily to go through the firebreak, and would thus certainly create the impression that the United States were willing to do so. 'It may be sound national policy: nonetheless, it represents the deliberate exclusion, though a self-denying ordinance, of a whole range of options.'[60]

From the foregoing survey of literature, it would seem that the critics identified that there could be and indeed there should be several dimensions and perceptions beneath the Assured Destruction. Scholars pay most attention to the inflexibility to translate power into policy whereas others stick their coin virtually on the threatened deprivation of city-hostages as an immoral act. Apart from the demand for a refinement and a revision, the dominant strain of comment stresses the folly of the Assured Destruction in light of the needed improvements in conceptual deterrence. The motive underlying the development of a new military posture was to provide for as many contingencies as possible with various nuclear options, and at the same time compatible with alternative resource-allocation demands.

Targeting: City-hostages or military establishments?

A number of targeting disadvantages develop as the two superpowers head to strength. Given that the target is likely to be a town or city, the strategic doctrine raises debilitating moral scruples. MAD is thus attended by all sorts of moral, political and alliance difficulties. A state genuinely holding a policy of this kind is likely to want to limit its commitment. The failure of the European countries to provide a respectable conventional force restrained part of the flexibility envisaged by McNamara's 'Flexible Response'.

Without going into great detail, the dilemma enshrined the targeting philosophy of either 'city-hostages or military establishments' and could well be explained by the 'conversion' of McNamara himself. In June 1962 he wrote,

> The United States has come to the conclusion that to the extent feasible, basic military strategy in a possible general nuclear war should be approached in much the same way that more conventional military operations have been regarded in the past. That is to say principal military objectives, in the event of a nuclear war, stemming from a major attack on the alliance, should be the destruction of the enemy's military forces, not of his civilian population . . . We are convinced that a general nuclear war target system is inadvisible, and if, despite all our efforts, nuclear war should occur, our best hopes lie in conducting a centrally controlled campaign against all of the enemy's vital nuclear capabilities, while retaining reserve forces, all centrally controlled.
>
> We know that the same forces which are targeted on ourselves are also targeted on our allies. Our own strategic

> retaliatory forces are prepared to respond against these forces, wherever they are and whatever their targets. This mission is assigned not only in fulfillment of our treaty commitments but also because the character of nuclear war compels it.[61]

In 1968 he wrote,

> Assured Destruction. . . is the clear and present ability to destroy the attacker as a viable twentieth century nation . . . a capability on our part to destroy, say, one-fifth to one-fourth of her population and one-half of her industrial capacity.[62]

Those who favor the 'city-hostages' targeting argue that if such a philosophy has to be actually executed, cities targeting is more credible in the sense that they are 'vulnerable' and because they are 'value'. Furthermore, before one could retaliate an 'unacceptable' damage on one's opponent after receiving the first-strike, most of the adversary's silos might already be empty. And furthermore since it is difficult to hedge those missiles out in the blue, it would be better to take the 'cities-hostages' targeting philosophy as inter-war deterrence and bargaining chips.

But given the adversary still has some nuclear weapons at his disposal, the belligerents become 'hostages' to one another. Such a relationship is well defined by Wolfgang K.H. Panofsky as 'the mutual-hostage relationship between America and Russia.'[63]

The strongest case against 'city-hostage' targeting philosophy is perhaps given by Donald G. Brennan. But what he actually prescribed is 'damage limitation' to the number of hostages either on the grounds of morality or feasibility, and is difficult to make out. This is precisely why I find it difficult to appreciate his argument, or to put it strongly I find it unconvincing in the apparent context of morality and humanity. What he actually proposed, I think, is 'graduate morality', which is not in the traditional sense 'morality of war'.[64] (But I am not interested in arguing the morality of war here.)

The same criticism is forcefully presented by Colin Gray who retorted, 'To repeat, if deterrence fails or is irrelevant, there is no point in executing civilian hostages. For the purpose of intra-war deterrence it would certainly be necessary to threaten grave impending costs unless . . . but killing people should have no part to play as a strategic objective.'[65]

The same moral indignation is, in fact, shared by most of the critics. To quote but a few examples. 'I cannot see how a city-targeting strategy can be reconciled with principle of the just employment of armed force even though to threaten destruction of millions of non-combatants is by no means as evil as would be their actual destruction'[66] concluded Arthur Lee Burns in Ethics and Deterrence; 'The Vulnerable Cities' as General Haywood S. Hansell[67] characterized the 20th century society.

What are the prescription/alternatives suggested by these critics? What really are the choices which the strategic systems permit in the event of nuclear firebreak?

In view of the destructiveness of the nuclear weapons, Fred Charles Ikle reacted strongly against the city-targeting philosophy saying, 'We believe, or we act as if we believe, that thanks to a certain international order, the existing arsenals of nuclear weapons with their almost incomprehensible destructiveness will never be used. Yet, this order is so constructed that it cannot move toward abolition of nuclear weapons.'[68] To him, the jargon of American Strategic Analysis works like a narcotic. It dulls our sense of moral outrage about the tragic confrontation of nuclear arsenals, primed and constantly perfected to unleash widespread genicide. In short, Fred Charles Ikle challenged the analysts of strategy for a better one than 'city-hostages'. But how can one do better? He is not explicit enough to give a guideline. No conclusive statements are made except to suggest a better control system in weaponry. Indeed, his conclusion sounds to me more like passing a death sentence than anything else: 'Despite the jargon of modern deterrence theory, ordinary Americans and Russians cannot escape the realization that their generation is destined to remain the chosen target of the nuclear forces on the other side.'[69]

In the same manner Colin Gray appears to have been in favor of a 'better strategy'. Better in what sense seems to depend on the ultimate effort of the strategists. The inherent follies of Assured Destruction thinking, argued Colin Gray,[70] stem especially from two sources: (i) many American strategists are still stranded on the 'intellectual doctrinal plateau' that was appropriate to the past and (ii) since the early 1960s a series of technological developments may offer great promise for a more sensible war-waging strategy (e.g. laser).

Herbert York's analysis arrived at the same conclusion stating that 'the overkill capacity in the present forces is so large that even substantial reductions would not do much to the threat hanging over the inhabitants of larger cities; most of them would be killed in the event of a breakdown of deterrence.'[71] What he prescribed is a non-nuclear strategy and arms control policy. He emphasized that the reduction of the current degree of overkill is both possible and essential.

Johan J. Holst[72] a less severe critic, prescribed an alternative by saying that the United States should be able to deter the Soviet Union by guaranteeing that it would be able to do at least as much damage to the Soviet Union as the Russians could inflict on the United States. Such a shift in outlook, Johan J. Holst maintained, might lead in the direction of a defensive emphasis posture. He has so much faith in 'such a posture' that it could also lead to renewed interest in controlled response and city-avoidance targeting in the event that a war should ever break out. However, the implication of this conclusion left much to be desired. All one could gather is he advocated for a more 'defensive posture'.

Meanwhile, Bruce Russett did make a deliberate attempt to contribute a 'theory of a countercombatant deterrent'. By counter-combatant he meant 'to include first all of those basic nuclear striking forces of the enemy that are currently targeted, and their immediate support facilities, missile silos, military airbases, submarine bases and nuclear submarines in port, air defense and ABM systems, and weapons-oriented atomic energy plants. More than that, it would include internal security forces, all military bases of any kind, and those transport facilities devoted primarily to the movement of troops and military supplies . . . so too, . . . weapons and military equipment-manufacturing plants would matter.'[73] Such options proposed by Bruce Russett in fact greatly enlarged the spectrum beneath the Assured Destrauction Capability. Of course, one should never forget the choice of options in Herman Kahn's ladder 'on thermonuclear war'.

The most absurd prescription is perhaps suggested by Raymond D. Gastil[74] who proposed a strategy that American leaders would renounce first use, even for the defense of Europe, and pledge never to directly attack civilians. And at the same time the United States would agree to respond to nuclear use anywhere in the world. No doubt if we take this supposition we are in the strategic deadlock.[75]

There is little doubt that in the face of physical reality, a relatively small investment in civil defense programs could save millions of lives in the event of nuclear war. Of course, if one could have everything, one could devote much for defense and much for offense at the same time. However, the U.S.S.R., if necessary, (whether they choose it or not is one thing, but that they could do it is another), increase the number of megatons delivered on target, change the compositoin of its warheads for more lethal and longer-lasting fall-out or redesign re-entry to penetrate deeply before exploding near blast shelter. Moreover, one could never be sure that the Russians would follow a city-avoiding strategy if the Americans chose such a 'self denying ordinance' as James R. Schlesinger put it.

In contrast to the non-nuclear policy approach, or to the 'cry in the wilderness for change', James R. Schlesinger described the then existing international order, and rationalized the range of options and the United States' weaponry systems. He concluded that the principle rationale behind Assured Destruction reflected certain recently developed notions of Arms Control. Such notions, of course, fell within the province of the decision-makers to adopt a strategy of measured response to any Soviet buildup with the long-term objective of preserving United States Assured Destruction Capabilities. 'I conclude the consequences of the narrowing range of acceptablility to be "harmful".'[76] In short, one feels that James R. Schlesinger did not renounce the deterrent principle inherent in the Assured Destruction. What he was dissatisfied with was the lack of options within the Assured Destruction doctrine.

Weaponry: Security through limitation--Possibility or Illusion?

The overriding concern in our world today is what to do with the nuclear weapons. The basic question is: do nuclear weapons have political utility? If the answer is negative, then one only needs to have an Assured Destruction capability to deter; this would be enough. If the answer is positive, then two questions arise: (i) can a nation gain a diplomatic advantage because it has a numerical edge in offensive nuclear weapons? (ii) is it useful to have certain kinds and numbers of weapons so that if deterrence fails a nation has a war-fighting or damage-limiting capability?[77] No one could ever give definite answers to these questions. SALT, in its basic sense, is part of the process to redefine the problem of nuclear weaponry in the 'era of negotiations', and to seek security through limitations, and to bring in line the American nuclear strategic doctrine with the whole pattern of international politics of detente.

To put it another way, as Thomas W. Wolfe phrased it, the period of attempted postwar containment of the Soviet Union has come to an end, but the rules of engagement under which the Soviet Union and the United States will look after their overlapping global interests are 'still in the process of being defined'.[78]

The rules of engagement indeed had already been written more than fifteen years ago by Thomas Schelling who suggested the notion of a 'strategy of conflict', calling for the development of a theory of international strategy. The 'strategy of conflict' is not concerned with the 'efficient application of force, but with the exploitation of potential force'. Conflicts in Schelling's conceptualization are not strictly zero-sum, win-or-lose situations. Both sides in a conflict may, in fact, gain. In such situations decisions are made to influence the other party's perception of how we will respond to a given act. It is the implementation of these rules that the two superpowers solve their strategy of conflict in the context of detente and in the Strategic Arms Limitation Talks.

Given the inflexibility of MAD as a strategic doctrine, the American society turns inward-looking; the apparent parity achieved by the Soviet; the expensive cost of engaging in a new arms race, the Nixon Administration took steps to remedy the situation by realpolitik. The approach is to negotiate with strength within the detente process to provide flexibility for its foreign objectives especially for the alliance obligation. However, the strategy of conflict - the exploitation of potential force - does not apply to the United States only. From the Test Ban Treaty to SALT I, the political antagonism between the two superpowers always remain. Arbatov explained in great detail the concept of peaceful coexistence as a "strategic principle of the revolutionary struggle" in his book on Ideological Struggle in Contemporary International Relations. He summed up the situation explicitly:

> No matter how successfully the process of normalization and of detente may develop, relations with the United States and other capitalist countries is in what forms will remain essentially based on struggle ...The question is in what forms will this struggle be fought?[79]

The best analogy is perhaps provided by James E. Dougherty. The long period of strategic weapons competition between the two superpowers is compared to a struggle between wolves:

> We have now arrived at that very moment in the contest when one of the wolves, not at all beaten yet feeling that he has had enough, temporarily and tenuously bares his neck while the adversary moves menacingly as if to overpower him. Those who study the behaviour of wolves explain that in this situation the onrusher, instead of moving in for the kill, can be expected to hesitate in a species-survival response that assures continued life to both.[80]

The two superpowers in this situation have no alternative but to watch each other carefully at this critical moment. They are still rivals, no doubt, each seeking foreign policy goals that are not perceived as compatible with those of the other. It is here in this context that nuclear weapons play the part of translating "power into policy" or in Schlesinger's phrase, "the strategy of conflict."

To regulate the behavior of 'wolves' in international politics, the superpowers began to talk in terms of Arms Limitation. However, while the Americans engaged themselves over the debate on whether to opt for a more offensive or a stronger defensive system, the Soviet Union was busy 'catching-up' (1965-1970) at least in terms of land-based missiles, involved improvements in both quality and quantity. The Soviet strategists got rid of this problem by recognizing that 'deterrence' unsupported by 'defense' was both dangerous and unstable, at the same time, the offensive-defensive mix or balance must have great advantages including that of 'surpassing' American Assured Destruction capabilities--'by-passing' as John Erickson[81] put it.

What actually happened was that the American Assured Destruction capability consisted of a triad of sufficient nuclear delivery forces--land-based ICBMs, submarine-launched ballistic missiles (SCBMs), and long-range heavy bombers. But from 1966 to 1967, the Soviet Union leaped from 338 ICBM launchers to 722--more than doubling the strength within one year; and then by 1968 to 902, and by the end of 1969 the Soviet Union had already possessed 1198. As the Sipri Yearbook (1974) pointed out by 1971 (mid-year figures) the total number of ICBMs the United States had was 1054 while the Soviet enjoyed a numerical advantage of 1527. Moreover, it became apparent at the beginning of 1970 that the SS-11 was being supplied with a more accurate warhead and was also being deployed within the medium-range launching systems in the south western areas of the Soviet Union, thereby diversifying Soviet capabilities both against Europe and the United States.[82] The SS-13, a solid-fuel missile comparable to the American Minuteman, has not received prominence as the SS-9 did when it came into service in 1965 and is capable of delivering a payload of 13,500 pounds, and in its first configuration a warhead of 25 MT on the United States. No doubt as Dr. Wohlstetter in the Sub-committee on National Security and International Operations Committee on Government Operations United States Senate: Planning-Programming-Budgeting reported in 1969:

> 500 SS-9s equipped with either of these MIRV options--four 1-MT MIRVs or three 5-MT MIRVs--could destroy about 95 percent of Minutemen if the Russians use well-established techniques for reprogramming missiles to replace known failures. Using no reprogramming at all, the 1-MT MIRV force would destroy 92 or 93 percent of Minutemen. The ability of the 5-MT force to destroy 95 percent of Minutemen presumes only half the failures after launch are replaced--a figure well within the state of the art.
>
> Even limiting the use of information to missile malfunctions before or during launch, the 5-MT MIRV force would leave only 8 or 9 percent surviving. These numbers are intrinsically uncertain--sensitive especially to changing accuracy. Four hundred SS-9s with 1-MT MIRVs and accuracies better by only 250 feet would destroy more Minutemen that 500 with the accuracy expected in the early 1970s.[83]

The demanding concern over the emerging United States-U.S.S.R. strategic relationship was solemnly voiced by President Nixon in his foreign affairs report of February 25, 1971:

> By any standard, we believe the number of Soviet strategic forces now exceeds the level needed for deterrence. Even more important than the growth in numbers has been the change in the nature of the forces the U.S.S.R. chose to develop and deploy. These forces include systems--particularly the SS-9 ICBM with large multiple warheads--which, if further

> improved and deployed in sufficient numbers could be uniquely suitable for a first strike against our land-based deterrent forces. The design and growth of these forces lead inescapably to profound questions concerning the threats we will face in the future, and the inadequacy of our current strategic forces to meet the requirements of our security.[84]

As far as the United States was concerned, the capability even for Mutual Assured Destruction was now eroded in face of the initiation of the U.S.S.R. enlargement and improvement of its strategic nuclear forces. Of course, as John Erickson pointed out that even assuming the Soviet Command has considered the first strike option, the attack on the Minutemen still leaves a great deal of American nuclear power available. Anyhow, under these circumstances,[85] three options were open to the Nixon administration: to expand American strategic forces and/or strengthen strategic defenses in an effort to gain or to retain superiority (which is difficult because of the budgetary restraint and the low profile desired by the silent majority); to adopt a doctrine of firing on warning (which is contradictory to the 'peace' goal announced by the President himself); and to head off threats and maintain strategic stability through some understanding, formal or tacit, on the limitation of strategic armaments (the only alternative left). To say that the Soviet Union had no motives to gain is to forget the very mechanicism of international politics. The power and prestige of a state exist only vis-a-vis the other state(s). Against the background of 'peaceful coexistence' which could be better termed as 'peaceful competition', the Soviet too have its interests. Such interests are fully described by Thomas W. Wolfe and John Erickson[86] and examined in terms of economic considerations; strategic considerations; the Soviet 'military-industrial complex', and the Institutional and Bureaucratic considerations.

The SALT migh serve Soviet purposes. Even in the absence of an agreement, the talks confirm Soviet parity with the United States; SALT feed detente; and create doubts and tensions among the American allies. Pondering these problems, Leonard Beaton has speculated: 'In SALT, as in so many things, it is possible to suspect that the American objective is essentially technical and the Soviet objective is political.'[87]

Whatever the motives, they are not our concern here in discussion the American strategic doctrine, and what we are really interested in is the outcome of the SALT I on the United States nuclear strategic capability of the already highly dissatisfied Mutual Assured Destruction doctrine.

To simlify the complexity, Document No. 92-113 is enclosed in the Appendix which is the content and interpretation on the limitation of Anti-ballistic Missile Systems and the Interim Agreement.

According to this document,[88] we work out the balance sheet. The Soviet gains are:

(1) The SALT agreement freezes the strategic arms race in fixed, land-based ICBM and SLBM launchers resulting in American inferiority in numbers of launchers and in throw weight. It prevents the United States from improving the survivability of its land-based forces through defense. In short the freeze works best for the Soviet Union at or just over the peak of her strategic buildup with a far more comprehensive Civil defense program, leaving behind the United States which has not yet taken the decision to respond to that 'defensive' buildup. The option to offset such Soviet superiority in numbers of launchers is to develop ULMs (undersea long-range missile systems).

(2) The psychological dynamics create tension on allies of both sides. The Soviet strategic buildup impressed others that the U.S.S.R. had now seized the initiative from the United States; and Soviet Union leaders can now deal with the United States with greater confidence as peers or superiors.

(3) However, American technological superiority in MIRVs is not frozen, nor is there any restriction on the Soviets to develop or achieve such technological superiority. It can then be expected that the Soviets will divert their resources in this area.

(4) By the limit on ABMs to be deployed, the American technological lead in ABM development is frozen giving the Soviet Union an advantage because the Soviet Union has already deployed a national capital area (NCA) defense system. Meanwhile, deployment of an ABM around Washington depends on congressional appropriations and public support.

(5) The Soviet Union can achieve superiority in submarine-launched ballistic missiles by taking full advantage of the options provided by the executive agreements to convert obsolete SLBMs (G- and H-class submarines) into Polaris-type SLBMs. Moreover, an expanded Soviet SLBM capability requires substantial increases in United States' budget outlays for anti-submarine warfare (ASW). Worse still, SLBMs can be fired from the sea south of the United States' border--a direction from which the United States lacks adequate warning.

(6) Both the United States and the Soviet Union may feel compelled to allocate resources for a new AMB system. Such a move is already accepted by the Soviets, but as for the United States, everything is dependent on the Congress.

(7) The Soviet Union gained a breathing space after her vigorous strategic buildup and it allowed her to allocate her resources for domestic and political base.

United States gains are:

(1) SALT I left the United States with leads in strategic bombers and MIRV technology. The agreement may prove advantageous if the United States continues in its R & D program and production where she enjoys technological superiority, but it

may prove disadvantageous if the United States pauses while the Soviet Union plunges full steam ahead.

(2) The Soviet numerical buildup in strategic offensive weapons is slowed down for the time being.

(3) SALT may provide incentive in Europe to play a greater role in her own security. The limitation of the Soviet AMB systems leaves amply targets for European deterrent forces. But SALT also creates a greater sense of detente, real or perceived, that may make the Europeans have lesser incentives.

In short, it is conceivable that the Soviet Union seeks to utilize the records to confirm Soviet-American strategic parity, to freeze United States quantitative inferiority in those areas where it existed, and to allow the Soviet Union to maximize its potential in qualitative dimensions of strategic armaments.

To sum up, asymmetry exists in relation to the differing strategic calculi used by the parties. What part does the Soviet strategic buildup play in the 'translation of power into policy'? What are the political objectives, if any, in the 'strategy of conflict' to 'exploit the potential force' in the 'era of negotiations' by the Soviet Union? Could the United States maintain its already highly inflexible strategic doctrine of Mutual Assured Destruction capability as a credible nuclear deterrence? Even if she could, could she fulfill her alliance obligation in face of the possible 'flexible' use in pursuing her political objectives gain by the Soviet Union in her strategic buildup? What would the Secretary of Defense of the United States have to offer?

FOOTNOTES (Chapter 3)

1 See Richard M. Nixon, "U.S. Strategy for the 1970s: A New Strategy for Peace," Washington, D.C.: U.S. Government Printing Office, Feb. 18, 1970; "U.S. Strategy for Peace: Building for Peace", Washington, D.C." U.S. Government Printing Office, Feb. 1971; and "U.S. Strategy for Peace: the Emerging Structure of Peace," Washington D.C., U.S. Government Printing Office, Feb. 9, 1972.

2 U.S. Foreign Policy for the 1970s: A New Strategy for Peace. A report to Congress by Richard Nixon, Feb. 18, 1970, Washington, D.C.: Government Printing Office, 1970, 2.

3 Paul Seabury, "On Detente", Survey, Spring 1973, vol. 19, No. 2(87), 62.

4 Arthur M. Schlesinger, Jr., A Thousand Days, Houghton Mifflin, 1965, 904. Theodore C. Sorensen quotes Kennedy as saying, 'We are not wedded to a policy of hostility to Red China. I would hope that . . . the normalization of relations . . . between China and the West . . . would be brought about', which suggests that Kennedy anticipated Nixon in at least the projection of a policy of reapproachment with China. Theodore C. Sorensen, Kennedy, Harper and Row, 1965, 665-736.

5 Sorensen expounded this theme in great detail.

6 William F. Kaufmann, The McNamara Strategy, Harper and Row, 1964, 152-158.

7 Speech of October 7, 1966.

8 U.S. Foreign Policy for the 1970s: The Emerging Structure of Peace. A report to the Congress by Richard Nixon, President of the United States, Feb. 9, 1972, Washington D.C.: Government Printing Office, 1972, 3.

9 See RAND Paper P-3545: On Relating Non-technical Elements to System Studies.

10 Henry Kissinger White House Press Release, June, 15, 1972.

11 Robert S. McNamara, "Excerpts from 1965 Posture Statement," Smith and Johns, American Defense Policy, 89.

12 Colonel Richard L. Curl, "Strategic Doctrine in the Nuclear Age", *Strategic Review*, Winter, 1975, 51.

13 Herbert T. York, "Deterrence by Means of Mass Destruction," *Bulletin of Atomic Scientists*, March 1974, Vol. XXX, No. 3, 4.

14 Robert H. Kupperman, Robert M. Behr and Thomas P. Jones, Jr., "The Deterrence Continuum," *Orbis*, Vol. XVIII, Fall 1974, No. 3.

15 For greater detail read Osgood and Tucker, *Use of Force*.

16 Ray E. Jones, *Nuclear Deterrence: A Short Political Analysis*, London, Routledge and Kegan Paul, 1968. See Introduction, 8.

17 *Ibid*., 14.

18 See William W. Kaufmann, "The Requirements of Deterrence" in William W. Kaufmann (ed.), *Military Policy and National Security*, Princeton: Princeton University Press, 1965.

19 Brent Scowcroft, "Deterrence and Strategic Superiority", *Orbis*, Vol. XIII, No. 2, Summer, 1969, 437.

20 See Herbert T. York, "Deterrence by Means of Mass Destruction," *Bulletin of Atomic Scientists* March 1974, Vol. XXX, No. 3, 405.

21 *Ibid*., 4-5.

22 *Ibid*, 4-5.

23 Although Colin Gray in "Unsafe at any speed: A Critique of Stable Deterrence Doctrine" questioned the rationale of the MAD analysis, I borrow the statement to direct to all analysts. RWSI Journal, June 1973, 23.

24 Raymond D. Gastil, "Missile Defense and Strategic Doctrine", Chapter 3 in *Why ABM?* ed. by Johan J. Holst and William Schneider, Jr., a collection of papers by staff and fellow members of Hudson Institute, Pergamon Press, 1969, 48.

25 Brent Scowcroft, "Deterrence and Strategic Superiority," *Orbis*, Vol. XIII, No. 2, Summer 1969, 435-454.

26 See Wolfgange K.H. Panofsky, "The Mutual Hostages Relationship", *Foreign Affairs*, Oct. 1973, 111.

27 *Ibid.*, 111

28 Colin Gray, "Unsafe at any speed: A Critique of Stable Deterrence Doctrine", RWSI, 1973, June, 25.

29 Paul C. Warkne testified in the United States Senate Committee on Foreign Relations, Sub-Committee on Arms Control, International Law and Organization, Arms Control Implications of Current Defense Budget Hearings, 92nd Congress, 1st Session, Washington, D.C.: Government Printing Office, 1971, 214.

30 *Ibid.*, 23.

31 *Ibid.*

32 James R. Schlesinger, RAND Paper P-3881, "Arms Interactions and Arms Control", reprinted RAND P-5284, 26.

33 James R. Schlesinger, RAND Paper P-3287, Dec. 1965, "The Changing Environment in Systems Analysis" reprinted RAND P-5284, 35.

34 James R. Schlesinger, RAND Paper P-3316, Feb. 1966, "Organizational Structure and Planning," reprinted RAND P-5284, 67.

35 See Brent Scowcroft article, *ibid*.

36 *Ibid*., 435-454.

37 *Ibid*., 450-451.

38 See General Bruce K. Holloway, U.S.A.F., "Reflections on Nuclear Strategy and the Nixon Doctrine", *Strategic Review*, Spring 1973, 7-14.

39 *Ibid*., p. 9.

40 See Raymond D. Gastil article in *Why AMB*, *ibid*.

41 *Ibid*., 47.

42 See Donald G. Brennan, "New Thoughts on Missile Defense," *Bulletin of Atomic Scientists,* June 1967, 10-15.

43 Quoted by James R. Schlesinger in RAND Paper P-3881, reprinted in RAND P-5284, 27.

44 See Donald G. Brennan, "The Case for Population Defense," Chapter 5 in *Why ABM?*, 91-117, especially p. 103.

45 See Fred Charles Ikle's article, "Can Nuclear Deterrence Last Out the Century," *Foreign Affairs*, Jan. 1975, Vol. 51, No. 2.

46 *Ibid*., 269.

47 Colin Gray, *op. cit*., 25.

48 See Herbert York's article, "Deterrence by Means of Mass Destruction," March 1970, *Bulletin of Atomic Scientists*, Vol. XXX, No. 3.

49 *Ibid*., 5.

50 See Martin J. Bailey's article, "Deterrence, Assured Destruction, and Defense", _Orbis_, Vol. XVI, No. 3, Fall 1972.

51 James R. Schlesinger, RAND Paper P-3545, Feb. 1967, "On Relating Non-Technical Elements to System Study", reprinted RAND P-5284, 82.

52 See Brent Scowcroft's article, _ibid_.

53 Colin Gray, _Ibid_., 25.

54 See Raymond D. Gastil's article, _ibid_.

55 James R. Schlesinger, RAND Paper P-3316, Feb. 1966 "Organizational Structures and Planning" reprinted RAND P-5284, 67.

56 Thomas C. Schelling, _Arms and Influence_, New Haven: Yale University Press, 1967, 193.

56a Clark A. Murdock, _Defense Policy Foundation_, Albany State University of New York Press, 1974, 71.

57 J.S. Coffey, "The Anti-Ballistic Missile Debate" _Foreign Affairs_, 45, April 1965, 409.

58 General Bruce K. Holloway, _Ibid._, 9.

59 See Johan J. Holst, "Missile Defense, The Soviet Union and the Arms Race", Chapter 7, in _Why ABM?_, 145-186.

60 James R. Schlesinger, RAND Paper P-3316, Feb., 1966, reprinted RAND P-5284, 67.

61 Robert S. McNamara's Speech at Ann Arbor, Mich. 16 June 1962.

62 Statement of Secretary of Defense McNamara on the Fiscal Years 1969-1973. Defense Program and the Defense Budget. Quoted in Hedley Bull, "The Scope for Soviet-American Agreement in Soviet-American Relations and World Order: Arms Limitations and Policy", Adelphi Paper, No. 65, 5.

63 Wolfgang K.H. Panofsky, Foreign Affairs, Oct. 1973, 109.

64 See St. Thomas Aquinas on War.

65 Colin Gray, Ibid., 25.

66 Arthur Lee Burns, "Ethics and Deterrence (A New Balance without hostage Cities?)", Adelphi Papers, No. 69, July 1970, 13.

67 The term is used by General Haywood S. Hansell, Jr., Strategic Review, Summer 1973.

68 Fred Charles Ikle, "Can Nuclear Deterrence Last Out the Century?" Foreign Affairs, Jan. 1973, Vol. 51, No. 2, 267.

69 Ibid., 282.

70 Colin Gray, Ibid., 26.

71 Herbert York, Ibid., 9.

72 Read Johan J. Holst's article, Chapter 7 in Why ABM?

73 Bruce Russett, "A Countercombatant Deterrent? Feasibility, Morality and Arms Control" from The Military-Industrial Complex: A Reassessment Sam Sarkesian (ed.), Beverly Hills, CA: Sage Publications, Inc., 1972.

74 See Raymond D. Gastil article, "Missile Defense and Strategic Doctrine," Chapter 3 in Why ABM?

75 The term is taken from Nuclear War: The Impending Strategic Deadlock, Neville Brown, London: Pall Mall, 1964.

76 James Schlesinger, RAND Paper P-3813, May 1968 "Defense Planning and Budgeting: The Issue of Centralized Control", reprinted RAND P-5284, 117.

77 In a detailed discussion, Symposium Discussion 108. SALT: Implications for Arms Control in the 1970s," Kinter and Pfaltzgraff (eds.) 1973, University of Pittsburg Press.

78 Thomas W. Wolfe, "Soviet Interest in SALT, *Ibid.*, 21.

79 Quoted by Leopold Labedz, "International Scenes in the Seventies," Spring 1973 *Survey*, Vol. 19, No. 2 (87), 5.

80 James E. Dougherty, "SALT and the Future of International Politics", Kinter and Pfaltzgraff (eds.), 359.

81 John Erickson, *Strategic Review Supplement*, Spring, 1973, 42.

82 *Ibid.*, 42.

83 Sub-committee on National Security and International Operations/Committee on Government Operations U.S. Senate: Planning-Programming-Budgetary. Defense Analysis: Two examples, Washington, 1969, 19.

84 U.S. Foreign Policy for the 1970s: Building for Peace. A report to Congress by President Richard Nixon, Feb. 25, 1971, 168.

85 In a further discussion of the difficulties of achieving meaningful stragegic advantages, see J.I. Coffey, *Strategic Power and National Security*, Pittsburg, University of Pittsburg Press, 1971.

86 Of course there are hundreds and hundreds of articles and books on SALT I, but by far the best for understanding the Soviet interests, I perfer Thomas W. Wolfe and John Erickson.

87 Secondary and Almost-Nuclear Powers: How do

they affect Strategic Arms Limitation? An occasional paper of the Center for Policy Study, University of Chicago, 1970.

88 See Appendix for Document 92-113.

CHAPTER 4

SYMMETRY AND SELECTIVITY

I Prelude: Motives and Circumstances

Perhaps the greatest contribution of President Nixon lies in illuminating the problem of contemporary nuclear strategy--unveiling its nature (the dynamics of the arms race and likewise of arms control), providing insights (if Schlesinger's strategy could finally provide any), and pointing out the need for more flexibility (either in the case of achieving one's foreign policy objectives or in the case of rejecting the inflexibility and incompatibility embedded in Mutual Assured Destruction in tune with the era of detente). This very explicitness in search of a viable strategy enables at least all strategists to understand more fully the whys and wherefores of Schlesinger's strategy. The overriding concern could be seen as early as 1970 in President Nixon's first foreign policy report:

> Should a President, in the event of a nuclear attack, be left with the single option of ordering the mass destruction of enemy civilians, in the face of the certainty that it would be followed by the mass slaughter of Americans? Should the concept of Assured Destruction be narrowly defined and should it be the only measure of our ability to deter the variety of threats one may face?

In February 1971, he repeated the same earnest desire:

> I must not be--and my successors must not be--limited to the indiscriminate mass destruction of enemy civilians as the sole possible response to challenges. This is especially so when that response involves the likelihood of triggering nuclear attacks on our own population.[1]

In February 1972, he re-emphasized the same theme:

> Our forces must also be capable of flexible application. A simple Assured Destruction doctrine does not meet our present requirements for a flexible range of strategic options.[2]

In May 1973 he forecast:

> We could no longer be complacent about the strategic status quo merely because we could cause a certain level of destruction in response to an attack. We therefore began to develop a sounder and more flexible doctrine for Air forces that would provide other retaliatory options besides a direct attack on millions of people.[3]

In short, President Nixon had made it clear from the beginning of his administration that he wanted changes in United States' strategic policy. The asymmetry perceived by the United States as a result of SALT I, the redefinition of United States' global objective of negotiation by strength, coupled with the momentum of technological changes in the weaponry system invited changes in her nuclear doctrine.

The asymmetry perceived by the United States as a result of SALT I has been propounded in the preceding Chapter. In summary the SALT I accords granted the Soviet Union 1618 ICBMs compared to 1054 for the United States. The Soviets were allowed to have 740 SLBMs with the option to build up to 950 in exchange for 210 of their older ICBMs. Meanwhile, the United States was limited to 656 with the option to build up to 710 in exchange for 54 Titan II ICBMs. In short, as Dr. Arthur G.B. Metcalf puts it, 'the SALT I accords freeze the United States in a position of strategic nuclear inferiority to the extent of a four to one disadvantage in missile payload and at a three to two disadvantage in numbers of launchers.'[4] Of course, one could always argue that the total of United States' strategic forces is in fact much larger. Of the 1054 ICBMs, 1000 are Minuteman III, which can carry up to three warheads. The 496 of the 656 SLBMs are scheduled (as recorded in 1973 preceding Schlesinger's announcement) to become Poseidon missiles carrying up to 14 MIRVs (although they usually deploy 10). Moreover, critics could always argue that approximately 400 B-52s and 65 FB IIIs are completely ignored in many comparative tables of American and Russian strategic forces, notable in Nixon's first three foreign policy reports.

Anyway, within six months of the signing of SALT I and the Interim Agreement on strategic offensive forces, many of the assumptions held by the American officials that SALT I would put a brake on the arms race proved wrong. Within a year or so the situation changed drastically. On 17 August 1973, Secretary of Defense James R. Schlesinger, announced that the Soviet Union had conducted successful tests of MIRV, for their new SSX-18 and SSX-17 ICBM. It is simply a matter of time before the SS-17 and SS-18 MIRV systems will achieve the accuracy needed to destroy hard targets.[5]

The Schlesinger announcement (on 17 August 1973) confirmed the fear of the many critics on SALT. Senator Jackson, the foremost Senate critic, was much disturbed and annoyed by the numerical inequalities contained in the terms of the Interim Agreement on Strategic Arms. The Jackson Amendment was under discussion from 2 August till 14 September 1972. It called for no less than numerical equality in any future treaty on offensive arms and the right to withdraw in the event that United States' strategic forces were threatened by Soviet 'modernizing activities'. To sum up the criticism of SALT I could well be stated as quantified Assured Destruction doctrine and has been rejected explicitly by President Nixon (in his relative damage criterion and by his insistence upon a flexibility of application of strategic forces but the terms of SALT I serve to strengthen, indeed to enshrine, the idea of Assured Vulnerability that is central to Mutual Assured Destruction.[6]

The initial act of strategic requirements established by the Nixon administration to cope with the redefinition of her global objectives and negotiation by strength involved the concept of 'sufficiency.' The concept of sufficiency was in part a political concept and was not based solely on debatable assumptions regarding possible scenarios. Sufficiency, to be precise, embraces two meanings: in its narrow military sense it means enough force to inflict a level of damage on a potential aggressor sufficient to deter him from attacking. In its broader political sense, sufficiency means the maintenance of forces adequate to prevent the United States and her allies from being coerced. Moreover, four 'sufficiency criteria' were clearly identified in Secretary of Defense Melvin Laird's Annual Report for FY 1973:

> (1) Maintaining an adequate second-strike capability to deter an all-out surprise attack on our strategic forces.
>
> (2) Providing no incentive for the Soviet Union to strike the United States first in a crisis.
>
> (3) Preventing the Soviet Union from gaining the ability to cause considerably greater urban/industrial destruction than the United States could inflict on the Soviets in a nuclear war.
>
> (4) Defending against damage from small attacks or accidental launches.[7]

These four criteria were all related to the narrow military sense, leaving the broader sense of sufficiency hanging in the air. However, faced with a perceived potential threat to the sufficiency of the United States' strategic forces her strategic objectives were to be redefined in 1973 together with the implementation of new programs. The strategic objectives were then identified as:

(1) To deter all-out attack on the United States or its allies.

(2) To face any potential aggressor contemplating less than all-out attack with unacceptable risks.

(3) To maintain a stable political environment within which the threat of aggression or coercion against the United States or its allies is minimized.[8]

In brief, the new strategic doctrine envisaged by President Nixon should embrace the idea of 'sufficiency'. But what is sufficiency? Sufficiency in what? In options? In weaponry? In overall strategic equivalence or balance? Does sufficiency mean parity or superiority? (Parity is a fact and superiority could be a myth.) The idea of superiority was at once repudiated in Secretary Kissinger's words, 'What in the name of God is strategic superiority? What is the significance of it politically, militarily, operationally at these levels of numbers? What do you do with it?' At the same time what worried the decision makers was that the Russians might believe otherwise. The Russians are seen to be striving not only for numerical parity but also insurance. Does the desire to have more flexibility and more options seem to be the same as superiority to the eye of the beholder? All these questions depend on perception; yet the question of perception is important only to be 'sensitized and alert to', and it is unlikely to yield firm and stable conclusions.[9] However, one should admit that the publication of the sufficiency criteria at least clarified the misleading image that American policies and forces only envisioned massive retaliation against cities.

In brief, the new strategic doctrine envisaged by President Nixon should embrace the idea of 'sufficiency". But what is sufficiency? Sufficiency in what? In options? In weaponry? In overall strategic equivalence or balance? Does sufficiency mean parity or superiority? Parity is a fact and superiority could be a myth.) The idea of superiority was at once repudiated in Secretary Kissinger's words, 'What in the name of God is strategic superiority? What is the significance of it politically, militarily, operationally at these levels of numbers? What do you do with it?' At the same time what worried the decision-makers was that the Russians might believe otherwise. The Russians are seen to be striving not only for numerical parity but also insurance. Does the desire to have more flexibility and more options seem to be the same as superiority to the eye of the beholder? All these questions depend on perception; yet the question of perception is important only to be 'sensitized and alert to', and it is unlikely to yield firm and stable conclusions.[9] However, one should admit that the publication of the sufficiency criteria at least clarified the misleading image that American policies and forces only envisioned massive retaliation against cities.

It has often been argued that technology has a momentum of its own. In fact, 'it is increasingly commonplace to assert that doctrine follows developed technology, which in turn reflects the preferences of institutional actors.'[10] To elucidate, exercise of the option to procure certain types of weapons may only be seen as a compromise between the limitations imposed by technological constraints and foreign policy objectives of the adversary. Thus whether it is intended or not, and whether one likes to admit it or not, the United States strategic policies and forces are the instrument of her foreign policy objectives.

Conscious of the momentum and limitations of technological advance, social scientists and theorists continue to explore the relationship between the deterrent use of the strategic forces and the weight of political objectives ostensibly justifying that use. More than a decade ago massive retaliation was broadened to flexible response as a counter to all levels of aggression and more recently 'graduated deterrence' has been oriented to cover a variety of extrahomeland objectives.

To cut a long story short, 'given the essentially qualitative nature of the strategic arms race, given the long lead time (seven years on average) for weapon development, given an appropriate irreducible minimum of political hostility between two pre-eminent great powers, and given the nature of the negotiating process,[11] the declaration strategy is in fact a translation of all these. To put it more explicitly, a strategic doctrine is supposed to provide guidance to the armed forces so that they may be organized, postured, and deployed in an appropriate manner. But it could also be argued as Colin Gray put it --'A strategic doctrine tends to be favoured by a military service not because it is the most appropriate one for a particular perceived strategic and political contest, but because it ensures a major role for that service.'[12]

We will discuss later on whether the momentum has any effect on shaping the Schlesinger doctrine, but meanwhile we will examine the ongoing programs preceding Schlesinger's announcement.

In 1972[13] the implementation of new programs which were later further elucidated in 1973, were first outlined in Nixon's Report to Congress. It was proclaimed by Nixon that the strategic weapons programs were actually planned within the framework of the Soviet strategic developments, arms limitations, and the potential for technological change:

> We are therefore improving our ICBM force. Silos for Minuteman missiles are being hardened, and 550 Minuteman III missiles with multiple independently targeted warheads will be deployed by the

mid-1970s.

Development of a new strategic submarine, the Trident, has been undertaken to provide a highly survivably replacement for our current ballistic missile submarines.

We are developing a generation of submarine launched missiles with substantially greater range. With these new missiles our Trident and Poseidon submarines will be able to operate in a much larger ocean area while still within range of targets, and thus will be less vulnerable.

The survivability of B-52 bombers has been increased by decreasing the time required for take-off on warning of an attack and by developing new basing concepts. This will reduce the threat from the growing force of Soviet ballistic missile submarines.

We have also begun engineering development of the B-1 bomber as a potential replacement for the aging B-52. The B-1 would maintain our bomber force as an important element in air mix of retaliatory forces, providing assurance against technological breakthrough, complicating an enemy's offensive and defensive planning, and ensuring flexibility of response.

The ABM facility at Grand Forks, North Dakota, is being completed. This installation will give us operational ABM experience while directly enhancing the survivability of Minuteman ICBMs. We will

> also continue our planning for the Washington, D.C. ABM site in order to provide additional security for the major control of our forces.
>
> Similarly, we are improving facilities for command and communications to control our responses in crisis situations.[14]

The above cumulative impact of pressures and prescriptions exerted by the President's aspiration for changes in the United States' strategic doctrine to extend a more credible deterrence on United States' allies in the political setting of the 1970's Superpower dialogue, to meet the technological momentum and restraints, to satisfy the role-seeking military lobby and to remedy the asymmetry perceived by the United States as a result of SALT I, anticipated a change in the American strategic doctrine. To recast the argument in a better form: the political-military doctrine of a state has to serve multiple functions--both internal and external.

Thus, the motives for change derive from growing anxiety on the credibility of extended deterrence, of the dissatisfaction on a nuclear strategic doctrine emphasizing massive retaliation against urban targets when the U.S.S.R. virtually possess the same capability, of the fear that a 'recognized inferiority of American strategic weapons vis-a-vis the size of Soviet strategic arsenals perceived by both allies and protagonists might confer disadvantages to the American policy-making.

2 Essence: Symmetry and Selectivity

Without doubt, any changes or shift of emphasis on current strategic doctrine must rationalize within the framework of the existing policies, capabilities and posture in the light of perceived threats to its security, interests and values. Indeed, it is argued (in *Strategic*

Survey 1974) that at the time Dr. Schlesinger moved to the Department of Defense in 1973, an inter-agency study of strategic targeting options under the direction of the Department was virtually completed. The conclusion was that the American strike plan was very short of small-scale strategic options, and that the massvie strikes on urban centres should only be taken as a last resort.

Indeed, it was more explicitly written in an article by W. Beecher on August 5, 1972, in the New York Times in which he spelled out certain directions of the new 'strategy':

> Major war plans are being revised by the White House; decisions made to develop but not deploy bigger more accurate arms; Kissinger heads study; more options for President sought--more being soft pedalled to curb debate.
>
> After three years of publicly and privately insisting it did not intend to develop strategic warheads able to destroy "hard" Soviet military targets, the administration has decided to develop but not at this point to deploy--more accurate and powerful warheads for existing Minutemen and Poseidon missiles to give them a better capability against just such targets.[15]

One month later (on September 22, 1972), James R. Schlesinger, who was still the Chairman of the Atomic Energy Commission anticipated much of the present shape of the United States' defense posture which would later earn him the caption of the 'Schlesinger doctrine', which he finally crystalized in the dramatic announcement of March 1974.

In fact on January 10, 1974, Secretary of

Defense Schlesinger had admitted that there had taken place in 'retargeting some missiles under a new strategic concept.' He emphasized that such a retargeting is to make clear to the Soviet Union that the United States has a variety of strategic options and thus to impress upon the Soviet Union that if no permanent agreement on handling strategic arms was reached, the United States was prepared to match it at every step in any nuclear arms buildup. It was 'a change in the strategies of the United States with regard to the hypothetical employment of central strategic forces. A change in targeting strategy as it were.'[16] He went on to amplify the changing in targeting strategy:

> To a large extent the American doctrinal position has been wrapped around something called "Assured Destruction" which implies a tendency to target Soviet cities initially and massively and that it is the principle option that the President of the United States or the Command Authorities would have in the event of a possible recourse to strategic weapons . . .
>
> What I was referring to is a set of selective options against different sets of targets. We would not necessarily specify any particular set of targets. Military targets, whether silos or other military targets, are, of course, one of the possible target sets. But it is necessary to maintain a set of options which goes beyond the inherent attack--all-out attack--against enemy cities in the event of nuclear exchanges.

Moreover, due to the dramatic growth of the Soviet force capabilities, 'the range of circumstances in which an all-out strike against an opponent's cities can be contemplated has narrowed considerably and one wishes to have alternatives for the employment of strategic forces other than what would be a suicidal strike against the cities of the other side.'

He repeatedly expressed in this initial announcement concern that if no permanent agreement was reached, the Soviet Union by 1980 could obtain 'a major counter-force capability which we lack.' However, at the same time he indicated that the retargeting decision is 'distinct' from the strategic arms talks. Thus he explained that even if a permanent agreement could be reached, the United States intended to maintain the same counterforce ability. 'What we are seeking is to forestall the development of an asymmetrical situation that would be beneficial (only) to the Soviet Union,' he concluded.

Later, in a News Conference on January 24, Schlesinger attempted to explain his initial announcement which was then widely interpreted that the United States' defense policy was moving away from the prevailing concept of mutual deterrence based on retaliatory attacks on to military targets. The change, he said, does not mean a significant alteration in the present list of targets. The change, he said, does not mean a significant alteration in the present list of targets. Nor, he continued, does it necessarily imply an expansion or improvement in the present arsenal of nuclear weapons. In short, the key to understanding the modifications is to recognize that the 'emphasis is upon selectivity and flexibility; that does not necessarily involve what is referred to as major counterforce capabilities . . . The emphasis is on the selection of targets.' Rather, he continued to argue, the objective is to give the President 'greater flexibility and selectivity' to use the present system to attack either Soviet cities or military targets. These targets would be bases, airfields or missile concentrations.

Finally, he acknowledged that 'in order to have a strategy of selectivity and flexibility one must consciously adapt and adjust plans to doctrine.

That is a considerable change.'[17] Thus he concluded that what is needed is improved command-and-control abilities as well as conscious changes in thinking on strategy. He expressed hope that 'by changing our targeting doctrine and explicitly doing so in view of deficiencies that have been discovered in the pre-existing doctrine, we feel we are enhancing deterrence and consequently we reduce the risk of war.'

In his Annual Report to Congress on March 4, 1974--The Report of the Secretary of Defense James R. Schlesinger to the Congress in the FY 1975 Defense Budget and FY 1975-1979--James R. Schlesinger elaborated on the changes and the reasons for them. This time Secretary Schlesinger clarified the issue and dispelled the confusion that derived from his earlier announcement to journalists and observers as to what the new policies encompassed. He remarked that the United States' thinking about strategic deterrence had not kept pace with the evolution of the threat. He revitalized the strategic thinking by saying:

> I frankly doubt that our thinking about deterrence and its requirements has kept pace with the evolution of the threats. Much of what passes as current theory wears a somewhat dated air--with its origins in the strategic bombing campaigns of World War II and the nuclear technology of an earlier era when warheads were bigger and dirtier, delivery systems considerably less accurate, and forces much more vulnerable to surprise attack.

After transmitting his keynote, he provided reasons for his new strategy. He contended that the disparities is strategic force levels--existing and potential--are gathering momentum. He continued that 'the scope of the Soviet program as it has now emerged is far more comprehensive than estimated even a year ago . . . a truly massive effort.' He went on to assert that 'we do not propose to let an opponent threaten a major component of our forces without our being able to pose a comparable threat. We do not propose to let an enemy put us in a position where we are left with no more than a capability to hold his cities hostage after the first phase of a nuclear conflict. And certainly we do not propose to see an enemy threaten one or more of our allies with his nuclear capabilities in the expectation that we would lack the flexibility and resolve to strike back at his assets (and those of any countries supporting the threat) in such a way as to make his effort both high in cost and ultimately unsuccessful.'

After spelling out reasons for his new strategy, he initiated the central piece of Schlesinger strategy--selectivity and symmetry--as a response to counter the critics of the Mutual Assured Destruction, to provide flexibility for the President in case of a nuclear war, to tune with the atmosphere of detente, and to provide credibility of extended deterrence.

<u>The Schlesinger 'Doctrine':</u>

(A) Selectivity

He admitted that, contrary to many observations made, the United States has not dropped out the Mutual Assured Destruction option. On the contrary, the Mutual Assured Destruction capability remains the last resort or backdrop of nuclear deterrence. But such capability is neither sufficient nor reassuring:

> Such reassurances may bring solace to those who enjoy the simple but arcane calculations of assured destruction. But they are of no great comfort to policymakers who must face the actual decisions about the design and possible use of the strategic nuclear force. Not only must those in power consider the morality of threatening such

> terrible retribution on the Soviet people for some ill-defined transgression by their leaders; in the most practical terms, they must also question the prudence and plausibility of such a response when the enemy is able, even after some sort of first strike, to maintain the capability of destroying our cities . . . Since we ourselves find it difficult to believe that we would actually implement the threat of assured destruction in response to a limited attack on military targets that caused relatively few civilian casualties, there can be no certainty that, in a crisis, prospective opponents would be deterred from testing our resolve.
> Today, such a massive retaliation against cities, in response to anything less than an all-out attack on the United States and its cities, appears less and less credible.[18]

But deterrence may fail. The massive retaliation of the Assured Destruction capability could not be expected to deter minor aggression. J.C. Garnett pointed out in The Theory of Limited War that 'it is recognized that threats, even threats of severe punishment are not always successful in deterring.' Henry Kissinger had long stressed that the purpose of deterrence is that what we may do will prove as effective as what we will do. The words of Kaufmann who had written in the 1960s could surely help towards understanding the rationale behind the Schlesinger strategy of selectivity-- 'It is probably hopeless to expect that a single deterrent will cover the entire range of contingencies and still satisfy the criteria of credibility.'

Thus, the responses to the above abuses were announced in Schlesinger's Report:

> Flexibility of response is also essential because, despite our best efforts,

> we cannot guarantee that deterrence will never fail; . . .
> To the extent that we have selected response options--smaller and more precisely focused than in the past--we should be able to bring all but the largest nuclear conflicts to a more rapid conclusion before cities are struck. Damage may thus be limited and further escalation avoided.[19]

The choice of the selectivity of options would totally depend on the nature and the scale of the enemy's attack and his objectives. Schlesinger indicated the options included 'not only cities and silos, but also airfields, many other types of military installations, and a variety of other important assets that are not necessarily collocated with urban population. We already have a long list of such possible targets; now we are grouping them into operational plans . . . to the extent necessary, we are retargeting our forces accordingly.

To repudiate the extreme views of Assured Destruction he frankly admitted that certain options have long existed. He reported to the Muskie Subcommittee of the Senate Committee on Foreign Relations as follows:

Senator Muskie: How will this new doctrine change the actuality of our targeting?

Secretary Schlesinger: The change in targeting doctrine comes about in the following way. Of course, all our delivery vehicles are targeted against specific targets. The point that is different about the targeting doctrine that I have outlined to you is the emphasis on selectivity and flexibility. In the past we have had massive preplanned strikes in which one would be dumping literally thousands of weapons on the Soviet Union. Some of those strikes could to some extent be withheld from going directly against cities, but that was limited even then.

With massive strikes of that sort, it would be impossible to ascertain whether the purpose of a strategic strike was limited or not. It was virtually indistinguishable from an attack on cities. One would not have had blast damage in the cities, but one would have considerable fallout and the rest of it.

Senator Case: This is the part that I think has to be stressed because I think it is the heart of anything new you have brought into this thing, but, in fact, there isn't anything new because already we have targets, we have weapons targeted, individual targets that can be selected.

Secretary Schlesinger: Yes

Senator Case: It is just a matter of thinking about it, isn't it?

Secretary Schlesinger: Right. That is why I referred to it as targeting doctrine rather than the term retargeting that has been employed. It is the question of firing doctrine, and how you view the problem. You are quite right, Senator, to stress that aspect.[20]

(B) Symmetry

In addition to the targeting doctrine, Secretary Schlesinger approached the problem of 'imbalance' derived partly as a result of SALT I and partly as a result of Soviet technological momentum by insisting on the principle of symmetry. The treatment of the principle of symmetry is not as definite and absolute as the principle of selectivity. Selectivity is to create an objective, open, explicit, verifiable, self-correcting process. But symmetry is presumably negotiable and intercahngeable. In short, the principle of symmetry was itself a reaction to the comprehensive rationality of the Soviet strategic buildup. To illustrate, Secretary Schlesinger stated,

> As far as we can judge, moreover, the Soviets now seem determined to exploit the asymmetries in ICBMs, SLBMs, and payload we conceded to them at Moscow.

But symmetry on a weapon-to-weapon basis is difficult, so he elucidated by saying,

> None of this should be taken to mean that exact symmetry must exist between the two offensive forces. The United States is willing to tolerate the existence of asymmetries provided that, in an area of alleged parity, they do not all favour one party. But we are not prepared to accept a situation in which all the visible asymmetries point in one direction. And we know from experience that the Soviets are not prepared to do so either.

It is apparent that the demand for symmetry in this sense evolves on the 'political significance of crude static indices.' 'Static numbers' seems to imply overall, unrefined, inventories of missles, and the term itself seems to have certain political advantages. 'While this presumably does not mean that Mr. Schlesinger would initiate a really foolish strategic program just because some misinformed person might be impressed by it, the remark is nevertheless a measure of the tremendous weight which he attaches to political perception of the strategic balance.'[21]

In order to achieve the principle of symmetry and the flexibility of selectivity and without any necessary association of the two--a posture Secretary Schlesinger clearly likes to refer to as 'essential equivalence'--the 'Principle Features of the Proposed Posture' was defined much more broadly by James R. Schlesinger himself:

> A capability sufficiently large, diversified, and survivable so that it will provide us at all times with high confidence of riding out even a massive surprise attack and of penetrating enemy defenses, and with the ability to withhold an assured destruction reserve for

an extended period of time.

Sufficient warning to ensure the survival of our heavy bombers together with the bomb alarm systems and command-control capabilities required by our national command authorities to direct the employment of the strategic forces in a controlled, selective and restrained fashion.

The forces to execute a wide range of options in response to potential actions by an enemy including a capability for precise attacks on both soft and hard targets, while at the same time minimizing unintended collateral damage.

The avoidance of any combination of forces that could be taken, as an effort to acquire the ability to execute a first strike disarming attack against the USSR.

An offensive capability of such size and composition that all will perceive it as in overall balance with the strategic forces of any potential opponent.

Offensive and defensive capabilities and programs that conform with the provisions of current arms control agreements and at the same time facilitate the conclusion of more permanent treaties to control, and if possible, reduce the main nuclear arsenals.[22]

3 Evaluation and Assessment

Evaluation and assessment of the Schlesinger doctrine can hardly be other than premature. Favorable and unfavorable criticisms are likewise put forward and widely debated. The Secretary's

Report is heralded by some as the dawn of a new day--'At long last it seems as if the United States is to have a serviceable strategic doctrine. At long last it appears that planning for the military defense of the United States faces up to the necessity for fighting a war, should deterrence fail.'[23] Simultaneously, the Secretary's strategy is condemned by some as unnecessary--'The improved counterforce capability proposed by the administration is viewed as not only unnecessary and potentially costly but also likely to undercut the SALT negotiations and increase the risk of nuclear war.'[24]

Since there is no alternative to premature assessment, and since his greatness will be measured by the degree to which Secretary Schlesinger can translate his programs into reality, and since there is no universally accepted criteria to assess and to evaluate a strategic doctrine, an attempt is made in this study by presenting the six criteria for discussion, namely;

1. To what extent the Schlesinger 'doctrine' is a response to harmonize with the political relationship vis-a-vis the Soviet Union. How far would such a regulatory attempt be likely to be successful?
2. To what extent the Schlesinger 'doctrine' is a response to reassure and revalidate the extended deterrence vis-a-vis her allies. How far would such a reassurance be likely to be appreciated?
3. To what extent the Schlesinger 'doctrine' is a response to the pressures and to offer a responsible image to the domestic public at large. How far would such an image be likely to be welcomed?
4. To what extent the Schlesinger doctrine is a response to reinstate the logic of deterrence and to satisfy the critics of strategy with a better footing of deterrence. How far would such a deterrent logic be likely to be acceptable?
5. To what extent the Schlesinger 'doctrine' is a response to the technology offered by the technological lobby as thus to afford the feasi-

bility of a city-avoidance strategy. Is there a new breakthrough in technology resulting from Schlesinger's implementation?

6. To what extent the Schlesinger 'doctrine' is a response to the bureaucratic pressures of the role-seeking military to find continued validity for missions.

Introduction to Discussion[25]

Before proceeding to a discussion of Schlesinger's doctrine--selectivity and symmetry--based on the six criteria, it may be useful to recapture the control emphasis of this study: 'the root of a state's strategic doctrine is its orientation toward those challenge confronting the state that involve the use of threat to use force.'[26]

In fact a number of American academic strategists in the sixties, including Speier, Brodie, Horelick and Knorr, anticipated that strategic parity would in the long run tempt the Soviet Union to extract political gains by relying on the rationality of its opponents. Their concern is based on the psychological assumptions and political dimensions of the strategic equation. This concern is at the root of their recommendation that the United States should sustain a position of at least marginal superiority.

By the 1970s, parity between the two superpowers was more or less recognized. The early fear that the Soviet Union might exploit parity translated itself from prediction to reality.

Against this background we can see that selective targeting is an attempt to preserve strategic nuclear weapons as a positive political instrument to serve American foreign policy objectives. It is widely viewed as Ann Arbor revisited. The rationale is to provide 'the maintenance of forces adequate to prevent us and our allies from being coerced'. The targeting strategy itself is an outcome of the action-reaction process--'The Soviet Union now has the capability in its nuclear forces to undertake selective strikes against targets other than cities. This poses us an obligation, if we are to ensure the credibility of our strategic deterrent, that we must be certain that we have a comparable capability in our strategic systems and in our targeting doctrine.'

In considering Schlesinger's proposals for a targeting doctrine, the vital question is: Is it necessary?

Broadly speaking, criticisms have been levelled at Schlesinger's proposals that the United States have already had the option of destroying many Soviet military industrial targets including air bases, naval bases, petroleum refineries, and steel mills with weapons delivered by missiles with accuracies of half a kilometer or better, and if with air bursts, with little fallout. Milton Leitenberg[27] has a deep and perhaps a noble belief that specific options had already been considered in the early and middle 1950s:

1 Military and civil bases;
2 Naval bases, including aircraft carriers, submaries and ASW forces;
3 ICBM installations;
4 Air defense installations, such as surface-to-air-missiles (SAMS) air bases, and radar;
5 ICBM early warning radar installations;
6 Satellite space tracking installations;
7 Military command, control and communications centre;
8 Military depots and stockpiles, particularly nuclear weapons;
9 Strategic (nuclear) production facilities;
10 Civilian port facilities;
11 Petroleum refineries and petroleum storage centres;
12 Electric power producing facilities;
13 Railroad marshalling yards;
14 Troop concentrations, concentration points, troops in the field;
15 Urban areas and cities.

To illustrate this hypothesis, A.T. Hadley wrote in 1961 that in 1960 to 1962 the SAC and the Tactical Air Command (TAC) could already deliver 18,000 to 20,000 megatons against the USSR which could scarcely suit a countercity doctrine or option.[28] This hypothesis that the United States has had the targeting 'options' is further substantiated by E.L.M. Burns who quoted that Robert McNamara's June 16, 1962 speech at Ann Arbor, Michigan asserted that the counterforce message 'had been given privately to NATO councils and elsewhere' before its public announcement. This suspicion is further triggered by Schlesinger's early announcement and Report--'The shifting in targeting strategy, I should emphasize, does not mean that we are pointing missiles away from city targets to military targets. The emphasis is on the selection of targets under certain hypothetical events that might occur in the future.' Related to this statement presented in Houston, Texas by USAF Chief of Staff on March 25, 1974 in which it is clearly stated, 'Of course, we did not in the past exclude military installations in our targeting. But now we have made this a statement of national policy.'

There would probably be less confusion if one remembers what is a strategic doctrine all about? My analysis throughout this study led to the conclusion that strategy is 'to translate power into policy'. Through the formulation and dissemination of doctrine, the political and military leadership of a state hammers out an agreed set of broad views about the role of force in the international life of the country and communicates them throughout the national security community of the government, to the legislation, and to the public at large.[29] Moreover, as it has already been noted any shift of emphasis or any innovation of a strategic doctrine must rationalize the existing capabilities and posture. After all, planning a strategy against strategic uncertainty and insecurity cannot neglect the political influence of nuclear weaponry, its potential for blackmail and the caution which it induces on the other.

In short, a strategic doctrine must fulfill its dual functions--external and internal consumption.

Discussion

On the superpower level, the principle of selectivity harmonizes with the detente process with its city-avoiding strategy. It provides a better atmosphere for superpower dialogue. It reinstates the 'supreme importance in war' which is to attack the enemy's strategy rather than to attack his cities. In short, detente offers opportunities and risk.

SALT, the by-product of detente, are not talks, but are competition in the form of negotiations wherein advantages and exploitation are sought. History has often proved and periodically occasions emerge in which negotiations become inappropriate to prevailing realities. The story of SALT I could hardly be other than a 'good thing.' Peaceful negotiation is a continuation of war by other means. A rational defense policy must meet the imperatives of national security without the unnecessary sacrifice to feed the detente process. 'If the seeking of a viable alternative to a declaratory policy that would outdo the excesses of Attila the Hun is destabilizing --then so be it. Stability exists in the eye of the beholder, and that eye interprets what it sees according to a dominant set of strategic ideas. It is strategic doctrine that must be altered. If strategic wisdom is sacrificed on the altar of detente, it were better to change one's religion.'[30]

Although Schlesinger has continuously repeated that the options issue is not to be coupled to SALT and the sizing of strategic forces, this seems to be ambiguous and inconsistent. Whatever strategic quantitative advantage one nation may have as measured by numbers of warheads, deliverable megatons, missiles and the like can be offset at least in part, by the other's strategic qualitative superiority in the flexibility available for the use of its nuclear weapons.

Thinking of the long term, Secretary Schlesinger's principle of selectivity is a salutary exercise to draw up a war-fighting strategy. The greater the flexibility in its nuclear weaponry utility than its adversary is likely to be taken as more impressive and advantageous. Moreover, one has to appreciate that the prinicple of selectivity is Schlesinger's painful attempt to provide an independent posture from the mirror-image of Soviet reaction by separating the issue of selectivity from negotiation. To Schlesinger, in any negotiating process, each side would soon come to learn a great deal about the fears, calculations and motivations that form the adversary's security policies. He, at least in part, attempts to provide a nuclear strategy to guarantee the strategic insecurity.

To repudiate that more flexible selective attacks would somehow provoke a Soviet reaction thus in turn destabilizing the strategic balance, thereby fuelling an arms race, Secretary Schlesinger rebuked such stupidity by saying:

> If the nuclear balance is no longer delicate and if substantial force asymmetries are quite tolerable, then the kinds of changes I have been discussing here will neither perturb the balance nor stimulate an arms race. If, on the other hand, asymmetries do matter (despite the existence of some highly survivable forces) then the critics themselves should consider seriously what responses we should make to the major programs that the Soviets currently have underway to exploit their advantages in numbers of missiles and payload.[31]

Indeed, the phrase 'interaction' vis-a-vis Soviet Union has long been partly repudiated by

James R. Schlesinger while he was serving in RAND. Arms race, as arms control, has logic of its own. To him, accepted doctrine tends sharply to limit permissible action and moderately limits permissible views. At any one time there is a 'somehow stereotyped view of threat, of proper strategy, and of acceptable armaments.' Thus, he continued to argue, rather than arms interactions being a game of subtle move and countermove based on high sensitivity to the logical implications of opponents actions, 'it reflects the erratic and occasional reaction and overreaction to dramatic and shocking achievements by the opponent.'[32] He criticized the arms controllers on the ground that they tend 'to deal with those parts of a problem that they consider subject to manipulate,'[33] and to leave out those they could not manipulate.

Furthermore, our knowledge on the principle of selectivity has its infancy in Schlesinger's approach to strategy in the 1960s. 'In planning for the future the appearance of uncertainty at a given time is perhaps less interesting than the certainty that changes in objectives and strategies will take place over time.'[34] His strongest contention and preoccupation is 'to provide greater sophistication regarding the character of nuclear war. This reflects the perceived reduction in the probability of an all-out nuclear holocaust. A wide spectrum of 'hedges' is thus illustrated by developing a creative attitude towards diversification and sophistication. He introduced his argument by saying that 'we face a wide spectrum of threats, but we cannot tell which, if any, will actually materialize.' Thus, the argument leads almost 'automatically to advocacy of flexibility and open-mindedness . . . with its attributes of avoiding commitments to specific goals, willingness to abandon obsolete plans, and preservation of multiple options.'[35] Moreover, United States' strategic capabilities will be determined by the forces deployed by potential foes, primarily the Soviet Union, and 'since predictions of how rapidly the Soviets would deploy strategic capabilities have repeatedly turned away, it would seem the better course of wisdom not to precipitate such questions prematurely.'[36] To amplify the argument, planning too far in advance means the forfeiting of options. He asserted that the gulf between wars for which the Americans prepare and the wrong wars that they fight at the wrong time and in the wrong places brings back to the organizational image of the outside world and the nature of foes tends to ossification and to obsolence. It should be understandable by now while Kissinger was conducting the superpower dialogue, Schlesinger was busy preparing for more flexibility of options. Very roughly, he advocated two approaches of planning which lend themselves to our understanding of the rationale behind the principle of selectivity: Cooks-Tour Planning and Lewis-and-Clark Planning.

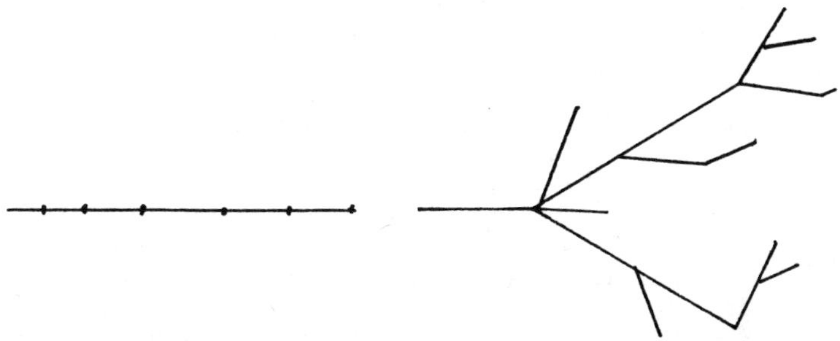

Cooks-Tour Planning Lewis & Clark Planning

To adjust ourselves to the Schlesinger approach to strategic planning, his whole piece is reproduced here:

Cooks-Tour Planning rests, implicitly

> or explicitly, on the supposition that that the future is sufficiently certain that we can chart a strategic course years in advance. In it, direction, speed, size of commitment, and achievement milestones (not decision points) are indicated with at least rough precision. By contrast, what may be termed Lewis-and-Clark Planning acknowledges that many alternatives, courses of action and forks in the road will appear, but their precise character and timing cannot be anticipated. Neither the size of commitment nor even the direction of movement should be stipulated too far in advance. At the end of a period one can retrospectively examine the paths pursued, which include many abandoned initiatives or experiments and many hard (and possible erroneous) choices. Only limited confidence could have been placed in advanced predictions regarding which options would be chosen, when the choices would be made, and how long alternative courses of action would be pursued before abandonment. Retrospectively one may map what has taken place, but the planning function is not to chart a precise course of action. Rather it is to prepare to cope with the uncertain terrain of the future, to note the signs in the environment that a decision point has been reached, and to respond in timely fashion.[37]

In short, Schlesinger argued that in spite of its messiness, the Lewis-and Clark Planning has its relative advantages in which it is conducive to facing uncertainties and hedging against uncertainties. Thus, one could feel the continutiy in Schlesinger's strategic planning finally materialized, articulated and implemented in the priniciple of selectivity.

It is often argued that MAD weakens the American nuclear umbrella and locks American strategic forces in inflexibility. Thus, 'the essence of Mr. Schlesinger's strategy for Europe remains the idea that in selective targetting the United States may discover some form of deterrence she could credibly wield against the Soviet Union when acting on Europe's behalf.' It is often believed that the United States must maintain a flexibile nuclear war fighting posture that could be employed in the defense of its allies, or else the process of "Finlandization" may emerge. Thus, the principle of selectivity may hope to preserve a credible deterrence by reducing the likelihood of nuclear holocaust for both superpowers.

One could in fact recall the intellectual orientation concerned with the allies at a news conference on 24 January when Secretary Schlesinger clarified the issue that 'the purpose to which we are altering that (targeting) doctrine is to enhance deterrence. It is necessary for the United States to maintain a worldwide military balance and a deterrent posture that offer no temptation to anybody, that preclude vacuums. By changing our targetting doctrine . . . we feel that we are enhancing deterrence and consequently, by enhancement of deterrence, we reduce the risk of war.'

However critics of Schlesinger's innovation may contemplate that the maintenance of tactical nuclear weapons and sizeable conventional forces in the local theaters and the linkages these provide to the strategic forces are said to be much more important than particular targeting plans or levels of counterforce capability.'[38] Within this framework of discussion 'perception speaks louder than realities.' It is easier to perceive the United States 'resolve' or 'will' in the form of United States' presence in Europe and a finger on the tactical nuclear weapons by the Europeans than a better selectivity in targeting.

Having considered the controversial interpretations, the principle of selectivity can be viewed as a logical extension of previous efforts to define a credible nuclear strategy for the allies by exploring the nature of protracted nuclear war. This does not imply that the allies have given a free ride. 'They should bear a fair share of the burden of keeping the peace and deterring major conflict.' However, whatever the 'practical outcome' of the principle of selectivity, it is important to move forward with efforts to improve the allies' conventional capability, particularly in the patterns of coping with the crises of instability. 'It must be repeatedly stressed that the deterrence of nuclear war depends not only on the adequacy of our strategic and tactical nuclear capabilities. It also rests heavily in the first instance on our ability to deter the outbreak of conventional conflict among the great powers. All of us recognize the theoretical utility of large-scale bolt-out-of-the-blue surprise nuclear attack scenarios for force planning purposes. Nonetheless, most of us would agree that the more likely first use of nuclear weapons would arise out of a setback at the conventional level of conflict. To keep the nuclear threshold high, we must therefore maintain strong conventional forces and work unceasingly to deter the outbreak of any major conflict.' (Speech by Schlesinger, Wednesday, 12).

Thus, on the whole the principle of selectivity would seem 'more justifiable--as well as more discreet--for Europeans to feel reassured, for, if one subscribes to the view that specific military options are of vital significance, the attempt to add flexible options is commendable in itself. If, on the contrary, one takes the line that Europe's real guarantee lies in the general impression of American commitment, such conscientious efforts to keep American strategy responsive to European contingencies must surely reinforce the necessary appearance of involvement.'

In the coures of this study, I have tried systematically to lay out a groundwork discussion on deterrence. In any given circumstances, the application of threatened sanctions must appear likely. Deterrence through Assured Destruction as a strategic policy at once received much criticism when it was first enunciated in the mid-1960s. Nations, it is argued, cannot be depended on to make unlimited sacrifices to achieve limited objectives. Critics of this concept believe it would be immoral and intolerable for the United States to exercise massive retaliation on Soviet population and industry. To sum up, as reviewed in the previous chapter, what appears in the criticisms of MAD is much talk of what deterrence is and of possible strategies that might be utilized in a deterrence situation; but there is little or no research into when certain strategies are used or what strategies tend to bring about what responses. What kinds of threats are those which leave no alternatives? Who would be likely to use these, and when do they tend to be used? Lacking in all criticisms is an unambiguous statement of whether, once the nuclear threshold is crossed, the level of damage which the deterrer is willing to maintain on behalf of the objective will vary with the value of the objective to the deterrer. In fact, the level of unacceptable damage is criticized but assumed constant equated with the maximum loss which would be absorbed. In a sense, the criticisms are based on any of three arguments: (1) An Assured Destruction policy implicitly discounts, to a quite unjustifiable degree, the possibility of an accidental or irrational attack by the USSR; if such an attack occurs and our response is one of massive retaliation, it will be catastrophic for the people of both nations, and others as well. (2) The threat of such retaliation is of doubtful, and diminishing credibility as a deterrent to limited Soviet attack particularly considering that, with the growth in Soviet strategic forces, there would be a virtual certainty of the United States' destruction should it visit such an attack on the USSR. (3) A massive retaliatory attack, whatever the provocation would be immoral, the more so in that most of those destroyed by it would have had little role in whatever event triggered the destruction.'[39] In short, MAD was not merely a system of deterrence, but deterrence in its most extreme, genocidal form.

In the light of these criticisms, the logical structure of Schlesinger's proposals may be discovered: can deterrence be more effective by his innovation?

What Secretary Schlesinger is proposing is the doctrine of Flexible and Selective Targeting options, popularly described by critics as counterforce, although Schlesinger is careful to distinguish between the two. At the Press Conference on 10 January, 1974, he defined the principle of selectivity: ' . . . that does not follow that we have a counterforce capability if one is able to destroy countervalue targets in that way. It depends upon the kind of military targets that you may be referring to. It is evident for example, that this large number of weapons in our stockpile provides us with the discriminating ability to go after certain classes of military targets--airfields and the like.' He further elucidated by reflecting that since the Russians have by this time a large second-strike capability it is necessary to have a change in United States' targeting doctrine. It is clear at the same time that the Soviet cities continue to be targetted. In fact, Schlesinger himself has never totally denounced the rationale of MAD. What he has argued as we have already noted is that MAD is neither sufficient nor reassuring.

Thus, it seems to suggest that Flexible Targeting Options means providing the President with the choice between Counterforce (missile silos) and/or Countervalue (people). In short, it is possible to argue that by the principle of selectivity the possibility of city avoidance depends crucially on the ability to attack other than and together with the countervalue capability, as well as depending on the declaratory policy and nature and scale of the adversary's attack. In terms of sheer words, in terms of capability, the principle of selectivity retains the Armageddon-like consequences of MAD and at the same time provides the options and flexibility within the parameter of MAD. The principle of selectivity, perhaps, is an answer and/or an attempt that lends greater weight to Daniel Ellsberg's words: 'It is a peculiarity of thermonuclear threats that they make cowards of practically everyone.'[40] One could even argue that Schlesinger recaptured the essence of Herman Kahn's "On Escalation" and McNamara's Ann Arbor speech by devising a possible war-fighting nuclear interwar deterrent strategy. Without the unreasonability and immorality of using people as the main hostages and bargaining chips for retaliation, the principle of selectivity heralded a 'Brave New World' and thus freed the academic strategists of their limbo.

In a period of search for military detente, the increased obsolescence of the notion of deterrence phrased in Ikle's words 'Could Deterrence Last Out of the Century?' or Schlesinger's forthright statement that deterrence has not kept pace with the evolution of threats, in fact, rests not merely on the changed character of the threat of national adversaries, but rests at the same time specifically on the attempt to explicate the political and moral objectives of military posture. 'A strategic doctrine is supposed to inform the deliberations of the legislators responsible for committing the resources and raising the military forces required and to mobilize a broad domestic consensus in support of the foreign and military policies of the government.' Under a policy of "No More Vietnams", the domestic conditions call for a strategy of no more civilian destruction.

Since 1968 unilateral and bilateral restraint without sacrificing genuine national needs or national security is the key advice that can lead to relevant reductions and refinements in defense posture. Military commitments and expenditure are scrutinized, to a degree that was rare in the past, by their moral as well as economic impacts on the domestic mood. George Liska's Imperial America summed up the picture of United States' mood by saying, 'the United States cannot effectively implement a global primacy . . . Unless it manages to insulate its society and economy from the traumatic impacts by each and every peripheral military involvement.'[41]

Goals and costs are examined in terms of national priorities. Various studies anticipate that the military budget would take a 'smaller share of the GNP in the 1970s' and with no significant worsening of 'international tensions.'[42] The competing claims on urban development, welfare, transportation, etc., would create a need for priorities 'to balance our aspirations with our resources.'[43]

The strongest dissenter is J.L. Clayton whose summation is self-evident, 'The rise of the military-industrial complex has unquestionably spurred the economy to new heights, but the benefits have passed most people by. The rich have been getting richer . . . whereas the poor have received only "crumbs from the table of Mars" . . . The whole system has been built upon the philosophy . . . that power equals security . . . Rarely have we considered non-military solutions. Only recently have we become aware . . . that our resources are limited, and that monies spent on defense necessarily reduce our options in solving pressing domestic needs.'[44]

A common feeling is the growing dissatisfaction within the United States society on national priorities. But this is not to say that chauvinistic nationalism has disappeared from the American scene, it simply is not as aggressive a motivating force as it used to be. Thus it could be argued that 'throughout the cold war it has been widely taken for granted that the domestic society would have to make do with whatever might be left after the claims of security, power and global mission had been satisfied. When such a presupposition is overtly as well as covertly challenged, as it increasingly is, and when challengers openly defy the imagery of chauvinistic nationals, and appeals to global mission, military claims on dollars compete with other claims on more necessary equal terms.'[45]

Thus, on the very first page of the Schlesinger Report he is quoting the psalmist to pacify the domestic mood to the effect that, 'where there is no vision the people perish.' In a reflex action, he continued, 'we can and must become increasingly competitive with potential adversaries in a more fundamental sense. We must not be forced out of the market--on land, at sea or in the air.'

Of course, a strategic doctrine must answer the basic question of what is the best we can do with what we get? The Force Posture and Capability that could enable the materialization of the Principle of Selectivity and the Principle of Symmetry (to be discussed later) must be coped within the Defense Secretary budgetary restraints. Secretary Schlesinger warned his domestic audience by emphasizing that 'more recently, illusions somewhat similar to those of the Fifties regarding deterrence have emerged about detente . . . if indeed we are to maintain a military posture second to none, there is no substitute for the hard, costly and unremitting effort required to keep up the nation's defense.'[46]

This is the way by which Schlesinger introduced his strategic doctrine: 'This is our first peacetime defense budget in a decade. It is therefore an appropriate time to consider best how to settle down for the long haul.' The success or failure of the Schlesinger doctrine will be very much dependent on the materialization of these programs that granted the capability and credibility of the principle of selectivity (and likewise the principle of symmetry).

The proposed programs forwarded by Schlesinger appear to fall into two categories: the short-term and the long-term programs. The short-term programs include procurement of more Minuteman III missiles, refinement of the existing guidance system of the Minuteman III to improve accuracy, a higher yield warhead for the Minuteman III, a general program to improve the accuracy of SLBMs, funds to flight-test Minuteman III with a large number of smaller re-entry vehicles. The long-term programs require considerable development time and the initial deployment would be in the 1980s. They include: development for a terminally guided 'manoeuvreable re-entry vehicle (MARV) for possible retrofit," the ICBMs and SLBMs, advanced Trident I SLBM designed for evasion of ABM interceptors, correction of the warhead's flightpath by further R & D, how the new MARV will work by either gas jets or by aerodynamic vanes, and the development of a new ICBM for the 1980s, which according to Schlesinger would give 'a very good capability against hard targets.'

Although this is still largely speculation, the proposed programs seem only to improve the 'marginal capability' of the existing MAD strategy. The way suggested to minimize the civilian fatalities is by careful target selection. Yet it seems incompatible with the amount of money spent to acquire a nuclear capability of destroying bridges, petroleum, depots, etc. Moreover, an actual nuclear exchange aimed at minimizing civilian casualties by the United States depends on what the Russians would do and how the Russians would react. Perhaps the most disquieting scene is that Russian strategic warheads have always been large.

Furthermore, it is interesting to examine the bureaucratic pressures and the role-seeking military on the Schlesinger doctrine. Two approaches are often used by decision-makers for weapon procurement. The weapon-system greatly utilized the foot-in-the-door technique by initiating a project as a feasibility study, without mentioning the total costs expected with the completion of the project well expressed in the statement that, 'once the camel had his nose in the tent, the full development program costed out to a billion dollars.'[47] The second approach was introduced to DOD by McNamara: the systems analysis or cost-effectiveness analysis. Indeed, the logic underlying systems analysis is viewed by many as that of radical rationality. Paul S. Hammond concluded:

> Theoretically the defense program budget, linked to systems analysis methods for the exposition of major defense policy issues, establishes common frames of reference for discussion throughout the military establishment and establishes what kind of data to extract and collect. This is in effect a "scientific" commitment to a rational organizational discourse.[48]

However the most cogent critic of cost-effectiveness analysis is James R. Schlesinger himself:

> Viewed abstractly, systems analysis implies rigorous thinking, hopefully quantitative, regarding the gains and the resource expenditures involved in a particular course of action--to insure that scarce resources are employed productively rather than wastefully. It is almost tantological therefore to state that systems analysis effectively employed will be beneficial.[49]

Nevertheless he was not against the systems analysis approach. What he was against at that time was that the defense intellectuals imbued with the sense of their 'rationality', pushed the systems analysis further than it reasonably could have been pushed. Coupled with his strategic Lewis-and-Clark approach and his attitude towards systems analysis, it is understandable that there are programs that apparently have not much influence in his strategic doctrine. Whether the advocacy of such programs is a result of the bureaucratic pressures and role-seeking military lobby is very difficult to confirm. But he has long been aware of the bureaucratic pressures. In February, 1967, he wrote:

> Decisions are nominally made by senior political figures who are harried, have

> insufficient time to study problems in detail, who are gripped by emotion of their youth or by prior experiences, and who are susceptible to claims made by subordinate groups which are couched in a way to appeal to their prejudices. Below them are a set of mutually jealous and warring bureaucratic groups, clamouring for resources and anxious to protect established preserves. To the extent that they are not closely watched, the subordinate bureaucratic groups will attempt to achieve their objectives quietly or even surreptitiously. Moreover, their capacity for resistance to high-level objectives enunciated from above, but to which they take exception, is breathtaking. Actual programs and allocative decisions will consequently diverge sharply from those that would be predicted on the assumption of a rational intelligence. Instead they will be strongly influenced by prejudice, sheer incompetency, and by outright fights, deviousness, and bootlegging within the bureaucracies.[50]

In this situation it was not surprising that the new strategic bomber, the B-1 (allocated $500 million for development funds) solved the problem for North American, which apparently has little to do with the targeting selectivity. To support this speculation, Schlesinger's Report is used to serve as evidence:

> If we are to continue to maintain an effective strategic bomber force through the 1980s and beyond, as I am convinced we should, we will eventually have to modernize that force. The principle improvements needed are: (i) faster airfield escape and greater protection against the effects of nuclear detonations in order to avoid destruction by SLBMs which might be launched in depressed trajectories from Soviet SSBNs, and (ii) a capability to fly at very low altitude at

> high subsonic speed in order to penetrate improved Soviet air defenses.

It is difficult to provide a clear cut answer as to whether such an improvement will maintain a greater selectivity, or keep an open option against uncertainties, or satisfy the role of the Air Force, or finally be looked upon as a pressure from the military-industrial complex. Anyhow, Schlesinger goes on, which may reflect such a posture is a result of the pressures exerted by the military-industrial complex and role-seeking military lobby,

> Although we have no evidence as yet that the Soviet Union is developing depressed trajectory SLBMs, or plans to operate its SSBNs close to our shores, or will undertake major new air defense programs at home, all of these capabilities will clearly be within its technological competency and economic capability.

Further speculation of the pressures could be supported by Boeing, the creator of B-29, B-47 and B-52, who received the B-1 'Avioici' (i.e. airborne electronic equipment) together with the continued overseeing of the production of Minuteman and the possibility in development of a new ICBM. Moreover, the F-15 Eagle and the F-14 Tomcat keep McDonnell Douglas and Grumman floating in the aerospace crisis since 1968. The package deal of high-low mix (i.e. expensive projects are balanced/mixed with cheap projects) expresses itself in the Narwhal submarine and the strategic cruise missile and the A-10 close Air Support aircraft. In fact, everybody gets something.

The Diego Garcia facility to maintain a presence in the Indian Ocean satisfies the role-seeking Navy lobby. The authorization of a DCP (Development Concept Paper) to define the characteristics of a proposed new carrier and the alternatives available will at least satisfy the Navy.

In fact, one could conclude that certain programs to maintain certain free structures is in fact an industrial battle. The maintenance of a particular military function supports a whole host of weapon systems, which in turn keep the industry alive. Without hesitation, one realizes that Grumman is a navy corporation; Boeing is an Air Force corporation. Competition between them, based on different 'capabilities', gets translated into intensive interservice rivalry ostensibly about differing weapon specifications. 'This in turn appears before the public as arguments about strategic doctrine.'[51]

In spite of the severe criticisms directed against Schlesinger's strategy, it is still held to be necessary. In partial summary, assessing the impact of the six criteria on the innovation of Schlesinger's Principle of Selectivity, one could come to the conclusion that it is necessary in a sense to provide a better atmosphere for superpower dialogue without sacrificing the risks of strategic uncertainties, to reassure her allies with a better deterrent capability, to satisfy the domestic-at-large by de-emphasizing the horror of MAD, to fulfill part of the criticism on MAD by developing greater flexibility, to battle out the bureaucratic pressures and the role-seeking military lobby by high-low mix projects.

Almost as important as my partial conclusion is the suggestion given by Milton Leitenberg, Charles Schwartz and Barry Carter in which they[52] all share the view that Schlesinger's Principle of Selectivity is nothing new. Selectivity of options has long existed. The evidence that they had sketched out is very convincing. In fact, Schlesinger himself never claimed that it is something new. What is new, perhaps, is for the first time the existing technological advancement fulfills the ambition of McNamara's Flexible Response. But as long as we could never get our fingers on the document of Single Integrated Operational Plan (SIOP), one could only make this supposition.

The Principle of Symmetry is perhaps prompted

by the desire to shift a doctrine to a more appropriate ground for SALT II. But 'doctrine should not be adjusted to accommodate that which is negotiable; rather, that what is negotiable should first be determined in domestic and interallied discussions over which is strategically desirable.'[53]

To understand the rationale of the Principle of Symmetry, one should examine the United States' objectives in SALT. Speaking on this issue, Kissinger said:

> Our objective with respect to the control of arms is as follows: obviously neither side should have military advantage as the result of any agreements that may be made. Secondly, neither side should be able to have a political advantage as a result of any aggreement that might be made. Thirdly, neither side should believe that such an advantage exists even if in reality it does not exist, because the perception is more important in many respects than the reality. And finally, neither side--no ally or other interested countries of either side should have such a perception.

Secretary Schlesinger's 1975 Defense Report talked on the same issue:

> There is also an important relationship between the political behavior of many leaders of other nations and what they percieve the strategic balance to be. By no means do all of them engage in the dynamic calculations about the interaction of Soviet and United States forces that have so affected our own judgements in the past. However, many do react to the static meausres of relative force size, number of warheads, equivalent megatonnage, and so forth. Hence to the degree that

> we wish to influence the perceptions of others, we must take appropriate steps (by their lights) in the design of strategic forces.

The same point was reported again in Secretary Schlesinger's 1976 Defense Report:

> The fourth requirement is for a range and magnitude of capabilities such that everyone--friend, foe and domestic audience alike--will perceive that we are equal of our strongest competitors. We should not take the chance that in this most hazardous of areas, misperceptions could lead to miscalculation, confrontation, and crisis.

It is against this background, we discuss the second part of Schlesinger's doctrine: the Principle of Symmetry.

On the superpower level, the Principle of Symmetry is in fact an appropriate response to safeguard the upgrading of Soviet strategic buildup. Secretary Schlesinger pointed out that since the 1972 Interim Agreement the Soviet had exploited the asymmetrical situation.

Judging from their strategic buildup, as recounted in the previous and the beginning of this chapter, this kind of near maximum exploitation of the permissive 'addition plus addition-by-replacement' provisions and deterrence plus coercive war-fighting capability, it is difficult to accept without any serious reservation on the part of United States, Secretary Schlesinger has made it clear that under some circumstances substantial improvements may become essential.

This is not the whole story. It also appears that the Soviet Union would continue to pursue and to exploit the detente diplomacy by acceleration of their strategic programs. Reacting imaginatively to this new international system that distinguished scholars like Stanley Hoffmann, Seyom Brown and Alastair Buchan are now arguing that world politics is fast becoming too complex a term for the strategic balance to have much meaning. None of these authors betray much alarm over the acknowledged shift in relative strategic power that has occurred in the past four years. As long as political significance could be attached to military unusable weaponry, it deems that the United States should offset the unfavorable numerical asymmetry that was implicitly or explicitly denied in SALT I.

Moreover, the attainment by the USSR of a credible strategic nuclear superiority can fulfill her political ambition by means of her strategic-coercive capability that increases her international prestige and affects the defense posture of the United States in certain areas. In what follows, distorted perceptions about the military balance may erode United States' deterrent credibility and thus undermine the nuclear guarantee to NATO. In short, the principle of symmetry provides a safeguard against 'wrong perception' and a guideline for SALT II.

However, a continued drive by the United States to match the Soviet in numbers of nuclear missiles weapons would be injurious and unwelcomed by the domestic-at-large. It also, in the end, contributed nothing to enhance the primary goal of national security. The magnetic appeal of arms control agreement remains clear. In spite of all the dissatisfaction felt by United States defense officials, and the criticisms forwarded by the academic strategists, the SALT II talks resumed in late February, 1974, between Schlesinger's press conferences in January and his Report in March.

In short, the Principle of Symmetry is necessary for external consumption. It is a necessity for a better superpower dialogue, and at the same time a guarantee against any uncertainties. To summarize, given the prevailing perception of a hostile Soviet buildup, a competitative situation of political utility of nuclear weaponry, creates the strongest incentives for carrying out a new strategic doctrine. As long as the matter of chrononlocial time could not be overcome or proved (i.e. evidence that technology has already made possible the selective targeting in time of McNamara) we welcome Secretary Schlesinger's innovation in its clear declaratory posture. The history of the Nuclear Age is dominated by the dialectical process of technological breakthrough and forces each country, in consequence, to recast its strategic doctrine. Schlesinger's innovation could partly be seen as a technological success and partly as his incentive and originality.

FOOTNOTES (Chapter 4)

1 U.S. Foreign Policy for the 1970s: Building for Peace. A report to Congress by Richard Nixon, President of the United States, Feb. 25, 1971, 58.

2 President Nixon's Foreign Policy Report, 1972, delivered to the US Congress, 9 Beb. 1972, 176.

3 US Foreign Policy for the 1970s: Shaping a Durable Peace. A report to Congress by Richard Nixon, President of the United States, May 3, 1973, 7.

4 Arthur G.B. Metcalf, "SALT II-Some Principles" Strategic Review, Summer, 1973, 7.

5 Taken from 1974 Posture Statement, 45-46.

6 Colin Gray, "Of bargaining Chips and building blocks: Arms Control and Defense Policy," International Journal, Spring 1973, Vol. XXVIII, No. 2, 277.

7 Melvin Laird, National Security Strategy of Realistic Deterrence, Annual Defense Department Report FY 1973, Washington DC 1972, 65.

8 Nixon Report to Congress, US Foreign Policy in the 1970s: Shaping a Durable Peace, May 3, 1973, 62.

9 For those who are interested in perception, see "Perceptions of the Military Balance: Models and Anecdotes", by Charles Wolf, Jr., March 1975, RAND, P-5402.

10 Taken from Colin Gray, "Arms Control and Defense Policy," International Journal, Spring 1973, Vol. XXVIII, No. 2. He in terms derived the argument from the testimony of Herbert York in Arms Control Implications of Current Defense Budget, 92nd Congress, 1st session.

11 *Ibid.*, 271. I borrowed the phrases from Colin Gray but used them in a different way.

12 *Ibid.*, 269.

13 1972 Report to Congress by President Nixon.

14 U.S. Foreign Policy for the 1970s: Shaping a Durable Peace by Richard Nixon, President of the United States, May 13, 1973, 63.

15 W. Beecher, "Major War Plans are being Revised by the White House," *New York Times*, August 5, 1972.

16 Remarks by Secretary of Defense, James R. Schlesinger, Overseas Writers Association Luncheon, International Club, Washington, DC, Jan. 10, 1974. Mimeo transcript, Office of the Assistant Secretary of Defense for Public Affairs, OASD/PA.

17 News Conference by Secretary of Defense, James R. Schlesinger at the Pentagon, Jan. 1974. Mimeo transcript, OASD 1 PA.

18 FY 75 DOD Report, 35, 37-38.

19 *Ibid.*, 38

20 US-USSR Strategic Policies, Report of Secretary of Defense, James R. Schlesinger to the Congress on the FY 1975 Defense Budget and FY 1975-1979 Defense Policy, P9.

21 Lawrence Martin, "Changes in American Strategic Doctrine--and Initial Interpretation," *Survival* July/August 1974, 159.

22 Hearing before a subcommittee of the Committee on Appropriations, House of Representatives, 93rd Congress, second session. Statement of Secretary of Defense, James R. Schlesinger, 110-111.

23 *Strategic Review*, Spring, 1974, United States Strategic Institute, 4.

24 Barry Carter, "Nuclear Strategy and Nuclear Weapons," *Scientific American*, May, 1974, Vol. 230, No. 5, 20.

25 To avoid repetition, the six criteria would be discussed under Selectivity and Symmetry respectively.

26 Inspired by Arnold L. Horelick, "Perspectives on the Study of Comparative Military Doctrines", May, 1973, RAND P-5026.

27 Milton Leitenberg, "Race to Oblivion," *Bulletin of Atomic Scientist*, Sept. 1974.

28 A.T. Hadley, *The Nation's Safety and Arms Control*, New York: Viking Press, 1961, 3, 4.

29 Arnold L. Horelick, "Perspectives", *op. cit*., 7.

30 Colin Gray, "Rethinking Nuclear Strategy", *Orbis* Winter 1974, Vol. XVII, No. 4, 1152.

31 FY 75 DOD Report, 39.

32 RAND P-5284, 25.

33 RAND P-5284, 31.

34 *Ibid*., 35.

35 *Ibid*., 55.

36 *Ibid*., 67.

37 *Ibid*., 58.

38 Ted Greenwood and Michael L. Nacht, "The New Nuclear Debate: Sense or Nonsense?", *Foreign Affairs*, 1974, July, 770.

39 G.W. Rathgens, "Flexible Response Options," *Orbis*, Vol. XVIII, Fall 1974, No. 3, 678.

40 Daniel Ellsberg, The Theory and Practice of Blackmail, Santa Monica, CA: RAND Corporation, P-3883, July 1968, 35.

41 George Liska, Imperial America, Baltimore, 1967, 106.

42 L.A. Lecht, Goals, Priorities and Dollars, New York: Free Press, 1966, 301.

43 Ibid., 84.

44 J.H. Clayton, (ed.), The Economic Impact of the Cold War, New York: Harcourt, Brace, and World, Inc., 1970, 281-282.

45 Harold Sprout and Margaret Sprout, "National Priorities: Demands, Resources and Dilemmas," World Politics, Vol. XXIV, Oct. 1971-July 1972, 311.

46 James R. Schlesinger, News Conference, Feb. 6, 1975.

47 David Novick, "Costing To-morrow's Weapon System," RAND Memorandum, 3170-PR, 1962, 2.

48 Paul Y. Hammond, "A Functional Analysis of Defense Department Decisionmaking in the McNamara Administration," American Political Science Review, 62, March, 1968, 62.

49 James R. Schlesinger, "System Analysis and the Political Process," RAND Paper 3464, Santa Monica, CA, RAND 1967, 1.

50 RNAD P-5284, 87-88.

51 Mary Kalder and Alexander Cockburn, "The Defense Confidence Game," New York Review of Books, June 13, 1974.

52 See Milton Leitenberg, "Race to Oblivion," and Charles Schwartz, "Helping the Pentagon Aim Right," and Barry Carter, "Nuclear Strategy and Nuclear Weapons."

53 Colin Gray, "Rethinking Nuclear Strategy," 1153.

CHAPTER 5

CONCLUSION

The purpose of this study has been to explore and assess the orientation and innovation of the Schlesinger doctrine in the light of the American historical strategic perspectives since 1945. Our analysis suggests that major shifts of emphasis in any strategic doctrine can be understood only as a means to provide a clue for a simple description of the complex, and a means to reconcile the poltical pressures of the time.

Strategy in fact bears the etchings and scribblings of the generation past. Those who are willing to scrutinize any innovation or in fact any strategic doctrine in a thoughtful and intelligent fashion must recognize the antecedents and sketch the elements of continuity.

With this negative conclusion, Schlesinger's doctrine--Selectivity and Symmetry--presumes a close relationship between McNamara's Ann Arbor speech. There is no reason for us to deny McNamara's originality in favor of James R. Schlesinger in their painful search for greater flexibility in the utility of nuclear weaponry. However, it is one thing to wonder and to be amazed at the originality of McNamara's Flexible Response; it is quite another to know to what extent McNamara had committed himself to implement his Flexible Response. His quick shift to Damage Limitation and Assured Destruction suggests that greater coherence was needed at the time between a declaratory posture and the technological performance. In short, Damage Limitation and Assured Destruction are only filling gaps awaiting a better viable nuclear strategic doctrine.

To some extent, the critics of Mutual Assured Destruction intensified the necessary precondition of Schlesinger's innovation and sharpened the cleavage between the then existing doctrine with its proper calculus--to translate power into policy. Similarly, MAD was incompatible and insufficiently congruent to the flows of international detente of the 1970s. Effective integration of an acceptable strategic doctrine to the political reality took in substantive interests.

Thus, within the pre-defined parameter of Mutual Assured Destruction, Schlesinger's doctrine --Selectivity and Symmetry--is an attempt to correlate strategy with reality. He clearly admitted, 'Since we ourselves find it difficult to believe that we would actually implement the threat of Assured Destruction in response to a limited attack on military targets and cause relatively few civilian casualties, there can be no certainty that, in a crisis, prospective opponents would be deterred from testing our resolve.'[1]

In short, both McNamara and Schlesinger attempted to seek flexibility within the nuclear parameter and, this, in turn, is an attempt to inject the dynamics of international politics to vitalize the apparent 'immobilism' of nuclear strategy. Perhaps the frustrations and dissatisfaction experienced ended up in different stories: technological restraints left McNamara's Flexible Response ineffective while James R. Schlesinger's selectivity and symmetry has been applauded successfully by the follow-up technological policy that could be implemented at this time.

In the light of the dynamics created by the changing strategic posture--Selectivity and Symmetry--and the external and internal tensions within the international political setting of the 1970s, Schlesinger's doctrine suggests a more flexible and innovative response to political and social demands from within and without. It is a panacea to some of the problems identified: to provide a better footing for SALT II, to design a better strategy for the allies, to demonstrate the 'resolve' and 'will' of the United States, to savour the new domestic theme of inward-looking American society, to build up a viable strategic doctrine for the 1980s. One could conclude that

Schlesinger's doctrine is a proper response to pressures from within and without. It embodies new initiatives, imaginative intellects, a flexible mental posture and realistic perceptions.

The danger remains, however. In emphasizing the manner as he did, Schlesinger's doctrine unfolds a new era, 'The Brave New World'. What I am worrying about differs from most of the academic strategists who fear that such a posture would lead to a head-on arms race between the two superpowers. My worry is that by its very nature the Schlesinger doctrine is a declaratory acceptance of a limited nuclear war. Is it true that our mentality has at last come to a point comparable to the pathos experienced before the two World Wars? The acceptance that nuclear war is inevitable? Could it be said that instead of trembling before nuclear weaponry makes us all cowards; Schlesinger now by his very doctrine turned and faced the dilemma that we have so tragically inherited.

A Brave New World? Or a final concession to an appalling inevitability?

FOOTNOTES (Chapter 5)

1 FY 75 DOD Report, 37-38.

APPENDIX

92D Congress 2d Session	HOUSE OF REPRESENTATIVES	Document No. 92-311

SALT AGREEMENTS

COMMUNICATION

from

THE PRESIDENT OF THE UNITED STATES

transmitting

Copies of the Treaty on the Limitation of Anti-ballistic Missiles Systems and the Interim Agreement on Certain Measures with respect to the Limitation of Strategic Offensive Arms signed in Moscow on May 26, 1972.

June 13, 1972 - Referred to the Committee on Foreign Affairs and ordered to be printed.

TREATY BETWEEN THE UNITED STATES OF AMERICA AND THE UNION OF SOVIET SOCIALIST REPUBLICS ON THE LIMITATION OF ANTI-BALLISTIC MISSILE SYSTEMS

The United States of America and the Union of Soviet Socialist Republics, hereinafter referred to as the Parties,

Proceeding from the premise that nuclear war would have devastating consequences for all mankind,

Considering that effective measures to limit anti-ballistic missile systems would be a substantial factor in curbing the race in strategic offensive arms and would lead to a decrease in the risk of outbreak of war involving nuclear weapons,

Proceeding from the premise that the limitation of anti-ballistic missile systems, as well as certain agreed measures with respect to the limitation of strategic offensive arms, would contribute to the creation of more favorable conditions for further negotiations on limiting strategic arms,

Mindful of their obligations under Article VI of the Treaty on the Non-Proliferation of Nuclear Weapons,

Declaring their intention to achieve at the earliest possible date the cessation of the nuclear arms race and to take effective measures toward reductions in strategic arms, nuclear disarmament, and general and complete disarmament,

Desiring to contribute to the relaxation of international tension and the strengthening of trust between States,

Have agreed as follows:

ARTICLE I

1 Each Party undertakes to limit anti-ballistic missile (ABM) systems and to adopt other measures in accordance with the provisions of this Treaty.

2 Each Party undertakes not to deploy ABM systems for a defense of the territory of its country and not to provide a base for such a defense, and not to deploy ABM systems for defense of an individual region except as provided for in Article III of this Treaty.

ARTICLE II

1 For the purposes of this Treaty an ABM system is a system to counter strategic ballistic missiles or their elements in flight trajectory, currently consisting of:

(a) ABM interceptor missiles, which are interceptor missiles constructed and deployed for an ABM role, or of a type tested in an ABM mode;

(b) ABM launchers, which are launchers constructed and deployed for launching ABM interceptor missiles; and

(c) ABM radars, which are radars constructed and deployed for an ABM role, or of a type tested in an ABM mode.

2 The ABM system components listed in paragraph 1 of this Article include those which are:

(a) operational;
(b) under construction;
(c) undergoing testing;
(d) undergoing overhaul, repair or conversion; or
(e) mothballed.

ARTICLE III

Each Party undertakes not to deploy ABM systems or their components except that:

(a) within one ABM system deployment area having a radius of one hundred and fifty kilometers and centered on the Party's national capital, a Party may deploy: (1) no more than one hundred ABM launchers, and no more than one hundred ABM interceptor missiles at launch sites, and (2) ABM radars within no more than six ABM radar complexes, the area of each complex being circular and having a diameter of no more than three kilometers; and

(b) within one ABM system deployment area having a radius of one hundred and fifty kilometers and containing ICBM silo launchers, a Party may deploy: (1) no more than one hundred ABM launchers and no more than one hundred ABM interceptor missiles at launch sites, (2) two large phased-array ABM radars comparable in potential to corresponding ABM radars operational or under construction on the date of signature of the Treaty in an ABM system deployment area containing ICBM silo launchers, and (3) no more than eighteen ABM radars each having a potential less than the potential of the smaller of the above-mentioned two large phased-array ABM radars.

ARTICLE IV

The limitations provided for in Article III shall not apply to ABM systems or their components used for development or testing, and located within current or additionally agreed test ranges. Each Party may have no more than a total of fifteen ABM launchers at test ranges.

ARTICLE V

1 Each Party undertakes not to develop, test, or deploy ABM systems or components which are sea-based, air-based, space-based, or mobile land-based.

2 Each Party undertakes not to develop, test, or deploy ABM launchers for launching more than one ABM interceptor missile at a time from each launcher, nor to modify deployed launchers to provide them with such a capability, nor to develop, test, or deploy automatic or semi-automatic or other similar systems for rapid reload of ABM launchers.

ARTICLE VI

To enhance assurance of the effectiveness of the limitations on ABM systems and their components provided by this Treaty each Party undertakes:

(a) not to give missiles, launchers, or radars, other than ABM interceptor missiles, ABM launchers, or ABM radars, capabilities to counter strategic ballistic missiles or their elements in flight trajectory, and not to test them in an ABM mode; and

(b) not to deploy in the future radars for early warning of strategic ballistic missile attack except at locations along the periphery of its national territory and oriented outward.

ARTICLE VII

Subject to the provisions of this Treaty, modernization and replacement of ABM systems or their components may be carried out.

ARTICLE VIII

ABM systems or their components in excess of the numbers or outside the areas specified in this Treaty, as well as ABM systems or their components prohibited by this Treaty, shall be destroyed or dismantled under agreed procedures within the shortest possible agreed period of time.

ARTICLE IX

To assure the viability and effectiveness of this Treaty, each Party undertakes not to transfer to other States, and not to deploy outside its national territory, ABM systems or their components limited by this Treaty.

ARTICLE X

Each Party undertakes not to assume any international obligations which would conflict with this Treaty.

ARTICLE XI

The Parties undertake to continue active negotiations for limitations on strategic offensive arms.

ARTICLE XII

1 For the purpose of providing assurance of compliance with the provisions of this Treaty, each Party shall use national technical means of verification at its disposal in a manner consistent with generally recognized principles of international law.

2 Each Party undertakes not to interfere with the national technical means of verification of the other Party operating in accordance with paragraph 1 of this Article.

3 Each Party undertakes not to use deliberate concealment measures which impede verification by national technical means of compliance with the provisions of this Treaty. This obligation shall not require changes in current construction, assembly, conversion, or overhaul practices.

ARTICLE XIII

1 To promote the objectives and implementation

of the provisions of this Treaty, the Parties shall establish promptly a Standing Consultative Commission, within the framework of which they will:

(a) consider questions concerning compliance with the obligations assumed and related situations which may be considered ambiguous;

(b) provide on a voluntary basis such information as either Party considers necessary to assure confidence in compliance with the obligations assumed;

(c) consider questions involving unintended interference with national technical means of verification;

(d) consider possible changes in the strategic situation which have a bearing on the provisions of this Treaty;

(e) agree upon procedures and dates for destruction or dismantling of ABM systems or their components in cases provided for by the provisions of this Treaty;

(f) consider, as appropriate, possible proposals for further increasing the viability of this Treaty, including proposals for amendments in accordance with the provisions of this Treaty;

(g) consider, as appropriate, proposals for further measures aimed at limiting strategic arms.

2 The Parties through consultation shall establish, and may amend as appropriate, Regulations for the Standing Consultative Commission governing procedures, composition and other relevant matters.

ARTICLE XIV

1 Each Party may propose amendments to this Treaty. Agreed amendments shall enter into force in accordance with the procedures governing the entry into force of this Treaty.

2 Five years after entry into force of this Treaty, and at five year intervals thereafter, the Parties shall together conduct a review of this Treaty.

ARTICLE XV

1 This Treaty shall be of unlimited duration.

2 Each Party shall, in exercising its national sovereignty, have the right to withdraw from this Treaty if it decides that extraordinary events related to the subject matter of this Treaty have jeopardized its supreme interests. It shall give notice of its decision to the other Party six months prior to withdrawal from the Treaty. Such notice shall include a statement of the extraordinary events the notifying Party regards as having jeopardized its supreme interests.

ARTICLE XVI

1 This Treaty shall be subject to ratification in accordance with the constitutional procedures of each Party. The Treaty shall enter into force on the day of the exchange of instruments of ratification.

2 This Treaty shall be registered pursuant to Article 102 of the Charter of the United Nations.

Done at Moscow on May 26, 1972, in two copies, each in the English and Russian languages, both texts being equally authentic.

For the United States of America:

RICHARD NIXON,
President of the United States of America.

For the Union of Soviet Socialist Republics:

L.I. BREZHNEV
General Secretary of the Central Committee of the CPSU.

INTERIM AGREEMENT BETWEEN THE UNITED STATES OF AMERICA AND THE UNION OF SOVIET SOCIALIST REPUBLICS ON CERTAIN MEASURES WITH RESPECT TO THE LIMITATION OF: STRATEGIC OFFENSIVE ARMS

The United States of America and the Union of

Soviet Socialist Republics, hereinafter referred to as the Parties,

Convinced that the Treaty on the Limitation of Anti-Ballistic Missile Systems and this Interim Agreement on Certain Measures with Respect to the Limitation of Strategic Offensive Arms will contribute to the creation of more favorable conditions for active negotiations on limiting strategic arms as well as to the relaxation of international tension and the strengthening of trust between States,

Taking into account the relationship between strategic offensive and defensive arms,

Mindful of their obligations under Article VI of the Treaty on the Non-Proliferation of Nuclear Weapons,

Have agreed as follows:

ARTICLE I

The Parties undertake not to start construction of additional fixed landbased intercontinental ballistic missile (ICBM) launchers after July 1, 1972.

ARTICLE II

The Parties undertake not to convert land-based launchers for light ICBMs, or for ICBMs of older types deployed prior to 1964, into land-based launchers for heavy ICBMs of types deployed after that time.

ARTICLE III

The Parties undertake to limit submarine-launched ballistic missile (SLBM) launchers and modern ballistic missile submarines to the numbers operational and under construction on the date of signature of this Interim Agreement, and in addition to launchers and submarines constructed under procedures established by the Parties as replacements for an equal number of ICBM launchers of older types deployed prior to 1964 or for launchers on older submarines.

ARTICLE IV

Subject to the provisions of this Interim Agreement, modernization and replacement of strategic offensive ballistic missiles and launchers covered by this Interim Agreement may be undertaken.

ARTICLE V

1 For the purpose of providing assurance of compliance with the provisions of this Interim Agreement, each Party shall use national technical means of verification at its disposal in a manner consistent with generally recognized principles of international law.

2 Each Party undertakes not to interfere with the national technical means of verification of the other Party operating in accordance with paragraph 1 of this Article.

3 Each Party undertakes not to use deliberate concealment measures which impede verification by national technical means of compliance with the provisions of this Interim Agreement. This obligation shall not require changes in current construction, assembly, conversion, or overhaul practices.

ARTICLE VI

To promote the objectives and implementation of the provisions of this Interim Agreement, the Parties shall use the Standing Consultative Commission established under Article XIII of the Treaty on the Limitation of Anti-Ballistic Missile Systems in accordance with the provisions of that Article.

ARTICLE VII

The Parties undertake to continue active negotiations for limitations on strategic offensive arms. The obligatwons provided for in this Interim Agreement shall not prejudice the scope or terms of the limitations on strategic offensive arms which may be worked out in the course of further negotiations.

ARTICLE VIII

1 This Interim Agreement shall enter into force upon exchange of written notices of acceptance by each Party, which exchange shall take place simultaneously with the exchange of instruments of ratification of the Treaty on the Limitation of Anti-Ballistic Missile Systems.

2 This Interim Agreement shall remain in force for a period of five years unless replaced earlier by an agreement on more complete measures limiting strategic offensive arms. It is the objective of the Parties to conduct active follow-on negotiations with the aim of concluding such an agreement as soon as possible.

3 Each Party shall, in exercising its national sovereignty, have the right to withdraw from this Interim Agreement if it decides that extraordinary events related to the subject matter of this Interim Agreement have jeopardized its supreme interests. It shall give notice of its decision to the other Party six months prior to withdrawal from this Interim Agreement. Such notice shall include a statement of the extraordinary events the notifying Party regards as having jeopardized its supreme interests.

Done at Moscow on May 26, 1972, in two copies, each in the English and Russian languages, both texts being equally authentic.

For the United States of America:

RICHARD NIXON
President of the United States of America.

For the Union of Soviet Socialist Republics:

L.I. BREZHNEV,
General Secretary of the Central Committee of the CPSU.

PROTOCOL TO THE INTERIM AGREEMENT BETWEEN THE UNITED STATES OF AMERICA AND THE UNION OF SOVIET SOCIALIST REPUBLICS ON CERTAIN MEASURES WITH RESPECT TO THE LIMITATION OF STRATEGIC OFFENSIVE ARMS

The United States of America and the Union of Soviet Socialist Republics, hereinafter referred to as the Parties,

Having agreed on certain limitations relating to submarine-launched ballistic missile launchers and modern ballistic missile submarines, and to replacements procedures, in the Interim Agreement,

Have agreed as follows:

The Parties understand that, under Article III of the Interim Agreement, for the period during which that Agreement remains in force:

The US may have no more than 710 ballistic missile launchers on submarines (SLBMs) and no more than 44 modern ballistic missile submarines. The Soviet Union may have no more than 950 ballistic missile launchers on submarines and no more than 62 modern ballistic missile submarines.

Additional ballistic missile launchers on submarines up to the above-mentioned levels, in the U.S. - over 656 ballistic missile launchers on nuclear-powered submarines, and in the U.S.S.R. - over 740 ballistic missile launchers on nuclear-powered submarines, operational and under construction, may become operational as replacements for equal numbers of ballistic missile launchers of older types deployed prior to 1964 or of ballistic missile launchers on older submarines.

The deployment of modern SLBMs on any submarine, regardless of type, will be counted against the total level of SLBMs permitted for the U.S. and the U.S.S.R.

This Protocol shall be considered an integral part of the Interim Agreement.

Done at Moscow this 26th day of May, 1972

For the United States of America:

RICHARD NIXON,
President of the United States of America.

For the Union of Soviet Socialist Republics:

L.I. BREZHNEV,
General Secretary of the Central Committee of the CPSU.

(Enclosure 3)

1. AGREED INTERPRETATIONS

(a) Initialed Statements

The texts of the statements set out below were agreed upon and initialed by the Heads of the Delegations on May 26, 1972.

ABM TREATY

(A)

The Parties understand that, in addition to the ABM radars which may be deployed in accordance with subparagraph (a) of Article III of the Treaty, those non-phased-array ABM radars operational on the date of signature of the Treaty within the ABM system deployment area for defense of the national capital may be retained.

(B)

The Parties understand that the potential (the product of mean emitted power in watts and antenna area in square meters) of the smaller of the two large phased-array ABM radars referred to in subparagraph (b) of Article III of the Treaty is considered for purposes of the Treaty to be three million.

(C)

The Parties understand that the center of the ABM system deployment area centered on the national capital and the center of the ABM system deployment area containing ICBM silo launchers for each Party shall be separated by no less than thirteen hundred kilometers.

(D)

The Parties agree not to deploy phased-array radars having a potential (the product of mean emitted power in watts and antenna area in square meters) exceeding three million, except for the purposes of tracking objects in outer space or for use as national technical means of verification.

(E)

In order to insure fulfilment of the obligation not to deploy ABM systems and their components except as provided in Article III of the Treaty, the Parties agree that in the event ABM systems based on other physical principles and including components capable of substituting for ABM interceptor missiles, ABM launchers, or ABM radars are created in the future, specific limitations on such systems and their components would be subject to discussion in accordance with Article XIII and agreement in accordance with Article XIV of the Treaty.

(F)

The Parties understand that Article V of the Treaty includes obligations not to develop, test or deploy ABM interceptor missiles for the delivery by each ABM interceptor missile or more than one independently guided warhead.

(G)

The Parties understand that Article IX of the Treaty includes the obligation of the US and the USSR not to provide to other States technical descriptions or blueprints specially worked out for the construction of ABM systems and their components limited by the Treaty.

INTERIM AGREEMENT

(H)

The Parties understand that land-based ICBM launchers referred to in the Interim Agreement are understood to be launchers for strategic ballistic missiles capable of ranges in excess of the shortest distance between the northeastern border of the continental U.S. and the northwestern border of the continental U.S.S.R.

(I)

The Parties understand that fixed land-based ICBM launchers under active construction as of the date of signature of the Interim Agreement may be completed.

(J)

The Parties understand that in the process of modernization and replacement the dimensions of land-based ICBM silo launchers will not be significantly increased.

(K)

The Parties understand that dismantling or destruction of ICBM launchers of older types deployed prior to 1964 and ballistic missile launchers on older submarines being replaced by new SLBM launchers on modern submarines will be initiated at the time of the beginning of sea trials of a replacement submarine, and will be completed in the shortest possible agreed period of time. Such dismantling or destruction, and timely notification thereof, will be accomplished under procedures to be agreed in the Standing Consultative Committee.

(L)

The Parties understand that during the period of the Interim Agreement there shall be no significant increase in the number of ICBM or SLBM test and training launchers, or in the number of such launchers for modern land-based heavy ICBMs. The Parties further understand that construction or conversion of ICBM launchers at test ranges shall be undertaken only for purposes of testing and training.

(b) Common Understandings

Common understanding of the Parties on the following matters was reached during the negotiations:

A. Increase in ICBM Silo Dimensions - Ambassador Smith made the following statement on May 26, 1972: "The Parties agree that the term 'significantly increased' means that an increase will not be greater than 10-15 percent of the present dimensions of land-based ICBM silo launchers".

Minister Semenov replied that this statement corresponded to the Soviet understanding.

B. Location of ICBM Defenses - The U.S. Delegation made the following statement on May 26, 1972: "Article III of the ABM Treaty provides for each side one ABM system deployment area centered on its national capital and one ABM system deployment area containing ICBM silo launchers. The two sides have registered agreement on the following statement: "The Parties understand that the center of the ABM system deployment area centered on the national capital and the center of the ABM system deployment area containing ICBM silo launchers for each Party shall be separated by no less than thirteen hundred kilometers. In this connection, the U.S. side notes that its ABM system deployment area for defense of ICBM silo launchers, located west of the Mississippi River, will be centered in the Grand Forks ICBM silo launcher deployment area." (See Initialed Statement (C).

C. ABM Test Ranges - The U.S. Delegation made the following statement on April 26, 1972: "Article IV of the ABM Treaty provides that 'the limitations provided for in Article III shall not apply to ABM systems or their components used for development or testing, and located within current or additionally agreed test ranges.' We believe it would be useful to assure that there is no misunderstanding as to current ABM test ranges. It is our understanding that ABM test ranges encompass the area within which ABM components are located for test purposes. The current U.S. ABM test ranges are at White Sands, New Mexico, and at Kwajalein Atoll, and the current Soviet ABM test range is near Sary Shagan in Kazakhstan. We consider that non-phased array radars of types used for range safety or instrumentation purposes may be located outside of ABM test ranges. We interpret the reference in Article IV to 'additionally agreed test ranges' to mean that ABM components will not be located at any other test ranges without prior agreement between our Governments that there will be such additional ABM test ranges."

On May 5, 1972, the Soviet Delegation stated that there was a common understanding on what ABM test ranges were, that the use of the types of non-ABM radars for range safety or instrumentation was not limited under the Treaty, that the reference in Article IV to "additionally agreed" test ranges was sufficiently clear, and that national means permitted identifying current test ranges.

D. Mobile ABM Systems - On January 28, 1972, the U.S. Delegation made the following statement: "Article V(1) of the Jount Draft Text of the ABM Treaty includes an undertaking not to develop, test, or deploy mobile land-based ABM systems and their components. On May 5, 1971, the U.S. side indicated that, in its view, a prohibition on deployment of mobile ABM systems and components would rule out the deployment of ABM launchers and radars which were not permanent fixed types. At that time, we asked for the Soviet view of this interpretation. Does the Soviet side agree with the U.S. side's interpretation put forward on May 5, 1971?"

On April 13, 1972, the Soviet Delegation said there is a general common understanding on this matter.

E. Standing Consultative Commission - Ambassador Smith made the following statement on May 23, 1972: "The United States proposes that the sides agree that, with regard to initial implementation of the ABM Treaty's Article XIII on the Standing Consultative Commission (SCC) and of the consultation Articles to the Interim Agreement on offensive arms and the Accidents Agreeemnt,*

---

*See Article 7 of Agreement to Reduce the Risk of Outbreak of Nuclear War between the United States of America and the Union of Soviet Socialist Republics, signed September 30, 1971.

agreement establishing the SCC will be worked out early in the follow-on SALT negotiations; until that is completed, the following arrangements will prevail: when SALT is in session, any consultation desired by either side under these Articles can be carried out by the two SALT Delegations; when SALT is not in session, _ad hoc_ arrangements for any desired consultations under these Articles may be made through diplomatic channels."

Minister Semenov replied that, on an _ad referendum_ basis, he could agree that the U.S. statement corresponded to the Soviet understanding.

F. _Standstill_ - On May 6, 1972, Minister Semenov made the following statement: "In an effort to accommodate the wishes of the U.S. side, the Soviet Delegation is prepared to proceed on the basis that the two sides will in fact observe the obligations of both the Interim Agreement and the ABM Treaty beginning from the date of signature of these two documents."

In reply, the U.S. Delegation made the following statement on May 20, 1972: "The U.S. agrees in principle with the Soviet statement made on May 6 concerning observance of obligations beginning from date of signature but we would like to make clear our understanding that this means that, pending ratification and acceptance, neither side would take any action prohibited by the agreements after they had entered into force. This understanding would continue to apply in the absence of notification by either signatory of its intention to proceed with ratification or ayproval."

The Soviet Delegation indicated agreement with the U.S. statement.

2. UNILATERAL STATEMENTS

(a) The following noteworthy unilateral statements were made during the negotiations by the United States Delegation: -

A. _Withdrawal from the ABM Treaty_

On May 9, 1972, Ambassador Smith made the following statement: "The U.S. Delegation has stressed the importance the U.S. Government attaches to achieving agreement on more complete limitations on strategic offensive arms, following agreement on an ABM Treaty and on an Interim Agreement on certain measures with respect to the limitation of strategic offensive arms. The U.S. Delegation believes that an objective of the follow-on negotiations should be to constrain and reduce on a long-term basis threats to the survivability of our respective strategic retaliatory forces. The USSR Delegation has also indicated that the objectives of SALT would remain unfulfilled without the achievement of an agreement providing for more complete limitations on strategic offensive arms. Both sides recognize that the initial agreements would be steps toward the achievement of more complete limitations on strategic arms. If an agreement providing for more complete strategic offensive arms limitations were not achieved within five years, U.S. supreme interests could be jeopardized. Should that occur, it would constitute a basis for withdrawal from the ABM Treaty. The U.S. does not wish to see such a situation occur, nor do we believe that the USSR does. It is because we wish to prevent such a situation that we emphasize the importance the U.S. Government attaches to achievement of more complete limitations on strategic offensive arms. The U.S. Executive will inform the Congress, in connection with Congressional consideration of the ABM Treaty and the Interim Agreement, of this statement of the U.S. position."

B. Land-Mobile ICBM Launchers

The U.S. Delegation made the following statement on May 20, 1972: "In connection with the important subject of land-mobile ICBM launchers, in the interest of concluding the Interim Agreement the U.S. Delegation now withdraws its proposal that Article I or an agreed statement explicitly prohibit the deployment of mobile land-based ICBM launchers. I have been instructed to inform you that, while agreeing to defer the question of limitation of operational land-mobile ICBM launchers to the subsequent negotiations on more complete limitations on strategic offensive arms, the U.S. would consider the deployment of operational land-mobile ICBM launchers during the period of the Interim Agreement as inconsistent with the objectives of that Agreement."

C. Covered Facilities

The U.S. Delegation made the following statement on May 20, 1972: "I wish to emphasize the importance that the United States attaches to the provisions of Article V, including in particular their application to fitting out or berthing submarines."

D. "Heavy" ICBM's

The U.S. Delegation made the following statement on May 26, 1972: "The U.S. Delegation regrets that the Soviet Delegation has not been willing to agree on a common definition of a heavy missile. Under these circumstances, the U.S. Delegation believes it necessary to state the following: The United States would consider any ICBM having a volume significantly greater than that of the largest light ICBM now operational on either side to be a heavy ICBM. The U.S. proceeds on the premise that the Soviet side will give due account to this consideration.";

E. Tested in ABM Mode

On April 7, 1972, the U.S. Delegation made the following statement: "Article II of the Joint Draft Text uses the term 'tested in an ABM mode', in defining ABM components, and Article VI includes certain obligations concerning such testing. We believe that the side should have a common understanding of this phrase. First, we would note that the testing provisions of the ABM Treaty are intended to apply to testing which occurs after the date of signature of the Treaty, and not to any testing which may have occurred in the past. Next, we would amplify the remarks we have made on this subject during the previous Helsinki phase by setting forth the objectives which govern the U.S. view on the subject, namely, while prohibiting testing of non-ABM components for ABM purposes: not to prevent testing of ABM components, and not to prevent testing of non-ABM components for non-ABM purposes. To clarify our interpretation of 'testing in an ABM mode', we note that we could consider a launcher missile or radar to be 'tested in an ABM mode' if, for example, any of the following events occur: (1) a launcher is used to launch an ABM interceptor missile, (2) an interceptor missile is flight tested against a target vehicle which has a flight trajectory with characteristics of a strategic ballistic missile flight trajectory, or is flight tested in conjunction with the test of an ABM interceptor missile or an ABM radar at the same test range, or is flight tested to an altitude inconsistent with interception of targets against which air defenses are deployed, (3) a radar makes measurements on a cooperative target vehicle of the kind referred to in item (2) above during the reentry portion of its trajectory or makes measurements in conjunction with the test of an ABM interceptor missile or an ABM radar at the same test range. Radars used for purposes such as range safety or instrumentation would be exempt from application of these criteria."

F. No-Transfer Article of ABM Treaty

On April 18, 1972, the U.S. Delegation made the following statement: "In regard to this Article (IX), I have a brief and I believe self-explanatory statement to make. The U.S. side wishes to make clear that the provisions of this Article do not set a precedent for whatever provision may be considered for a Treaty on Limiting Strategic Offensive Arms. The question of transfer of strategic offensive arms is a far more complex issue, which may require a different solution."

G. No Increase in Defense of Early Warning Radars

On July 28, 1970, the U.S. Delegation made the following statement: "Since Hen House radars (Soviet ballistic missile early warning radars) can detect and track ballistic missile warheads at great distances, they have a significant ABM potential.

Accordingly, the U.S. would regard any increase in the defenses of such radars by surface-to-air missiles as inconsistent with an agreement."

* * * * * * * * *

(b) The following noteworthy unilateral statement was made by the Delegation of the U.S.S.R. and is shown here with the U.S. reply:

On May 17, 1972, Minister Semenov made the following unilateral "Statement of the Soviet Side": "Taking into account that modern ballistic missile submarines are presently in the possession of not only the U.S., but also of its NATO allies, the Soviet Union agrees that for the period of effectiveness of the Interim "Freeze" Agreement the U.S. and its Nato allies have up to 50 such submarines with a total of up to 800 ballistic missile launchers thereon (including 41 U.S. submarines with 656 ballistic missile launchers). However, if during the period of effectiveness of the Agreement U.S. allies in NATO should increase the number of their modern submarines to exceed the numbers of submarines they would have operational or under construction on the date of signature of the Agreement, the Soviet Union will have the right to a corresponding increase in the number of its submarines. In the opinion of the Soviet side, the solution of the question of modern ballistic missile submarines provided for in the Interim Agreement only partially compensates for the strategic imbalance in the deployment of the nuclear-powered missile submarines of the USSR and the U.S. Therefore, the Soviet side believes that this whole question, and above all the questions of liquidating the American missile submarine bases outside the U.S., will be appropriately resolved in the course of follow-on negotiations."

On May 24, Ambassador Smith made the following reply to Minister Semenov: "The United States side has studied the 'statement made by the Soviet side' of May 17 concerning compensation for submarine basing and SLBM submarines belonging to third countries. The United States does not accept the validity of the considerations in that statement."

On May 26 Minister Semenov repeated the unilateral statement made on May 24. Ambassador Smith also repeated the U.S. rejection on May 26.

BIBLIOGRAPHY

Government and Official Publications

Dulles, John Foster, "A Policy for Peace Insurance", address at International House, May 16, 1950, DSB, XXII, May 29, 1950.

"The Interdependence of Independence", address at Sesquicentennial Fourth of July Celebration, DSB, XXIII, July 17, 1950.

"The Challenge of Today", address at University of Arizona, May 30, 1951, DSB, XXIV, June 11, 1951.

Statement before Senate Foreign Relations Committee, Jan. 21, 1952, DSB, XXVI, Feb. 4, 1952.

Transcript of "Meet the Press", NBC Television, Feb. 10, 1952.

"Morals and Power", address at National War College, June 16, 1953, DSB, XXVIII, June 29, 1953.

"The Evolution of Foreign Policy", address before the Council on Foreign Relations, Jan. 12, 1954, DSB, XXX, Jan. 25, 1954.

"Report to the President", May 17, 1955, Dept. of State Publication.

Address on United States Foreign Policy, June 24, 1955, DSB, XXXIII, July 4, 1955.

"Entering the Second Decade", Sept. 22, 1955, DSB, XXXIII, Oct. 3, 1955.

"The Challenge of Change", statement before the Senate Foreign Relations Committee, June 6, 1958, DSB, XXXVIII, June 23, 1958.

"Principles and Policies in a Changing World", November 18, 1958, DSB, XXXIX, Dec. 8, 1958.

President Johnson, Speech of October 7, 1966.

Kissinger, Henry, White House Press Release, June 15, 1972.

Laird, Melvin R., Annual Defense Dept. Report, FY 1971, statement before the House Armed Services Committee, Washington: Government Printing Office, 2/70.

McNamara, Robert S., Statement before Senate Sub-committee on Defense Appropriations, 18 April, 1961.

Statement before House Committee on Armed Services, 24, January, 1962.

Speech to American Bar Foundation, 17 February 1962.

Speech at Univeristy of Michigan, Ann Arbor, June 16, 1962.

Statement before House Armed Services Committee on Defense Budget, Jan. 30, 1963.

Statement before House Armed Services Committee on Defense Budget, Jan. 27, 1964.

Statement before House Armed Services Committee, Feb. 18, 1965.

Military Posture Statement to Joint Session of Congress and Armed Services Committee, Jan. 23, 1967.

Nixon, Richard, "United States Foreign Policy for the 1970s: Building for Peace", a report to Congress, Feb. 25, 1971.

"Foreign Policy Report 1972", Feb. 9, 1972.

"United States Foreign Policy for the 1970s: Shaping a Durable Peace", May 3, 1973.

Schlesinger, James R., America's Defense Posture, a statement by United States Secretary of Defense, Feb. 6, 1975.

Excerpts from the Annual Report to Congress of the United States Defense Department, Feb. 5, 1975.

A Press Conference on Nuclear Weapons, June 23, 1975.

Nuclear Weapons, an outline of United States Policy July 3, 1975.

Overseas Writers Association Luncheon, International Club, Washington DC, Jan. 10, 1974, Mimeo transcript, Office of the Assistant Secretary of Defense for Public Affairs, OASD/PA.

News Conference by Secretary of Defense, James R. Schlesinger, at the Pentagon, Jan., 1974, Mimeo transcript, OASD/PA.

Senate Hearings before the Commitee on Appropriations, Dept. of Defense Appropriations, 92nd Congress, second session, Fiscal Year 1973.

Senate Hearings before the Committee on Appropriations Dept. of Defense Appropriations, 93rd Congress, first session, Fiscal Year 1974.

Senate Hearings before the Committee on Appropriations, Dept. of Defense Approporiations, 93rd Congress, second session, Fiscal Year 1975.

Hearings before a Subcommittee on Appropriations House of Representatives, 93rd Congress, second session, Part I.

<u>Books and Articles by James R. Schlesinger</u>

James R. Schlesinger, "The Budget and National Security", <u>Current History</u>, July 1960.

James R. Schlesinger, "Of Pieties and Policy", Reporter, August 12, 1965.

"Strategic Consequences of Nuclear Proliferation", Reporter, Oct. 20, 1966.

"Nuclear Spread", Yale Review, October, 1967.

"Soft Factors in Systems Studies", Bulletin of Atomic Scientists, November 1968.

"Systems Analysis and the Political Process", Journal of Law and Economics, November 1968.

in Problems of National Strategy, Henry A. Kissinger (ed.), Praeger, 1965.

in Issues in Defense Economics, Roland N. McKean (ed.), Columbia University Press, 1967.

Books

Aron, Raymond, The Imperial Republic: The United States and the World 1945-73.

Bell, Coral, Negotiation from Strength: A Study in the Politics of Power, London: Chatto and Windus, 1962.

Brennan, Donald C. & Holst, Johan J., Ballistic Missile Defense: Two Views, Adelphi Paper No. 43.

Brodie, Bernard, Strategy and National Interests. Reflections for the Future, National Strategy Information Center, Inc., 1971.

Brown, Neville, Nuclear War, London: Pall Mall Press, 1964.

Brown, Seyom, The Faces of Power: Consistency and Change in United States Foreign Policy from Truman to Johnson, New York & London, Columbia University Press, 1968.

Burns, Arthur Lee, Ethics and Deterrence. A Nuclear Balance without Hostage Cities? Adelphi Papers, No. 69.

Chayes, Abram and Wiesner, Jerome B., ed. ABM, An Evaluation of the Decision to Deploy an Anti-ballistic Missile System, Signet Broadside, New York, The New American Library, 1969.

Clayton, J.H., ed., The Economic Impact of the Cold War, New York: Harcourt, Brace & World, Inc., 1970.

Ellsberg, Daniel, The Theory & Practice of Blackmail, Santa Monica, CA, The RAND Corporation P-3883, July 1968.

Enke, Stephen, Ed., Defense Management, Englewoods Cliffs, N.J.: Prentice-Hall, Inc., 1967.

Enthoven, Alain C. & Smith, K. Wayne, How Much Is Enough? Shaping the Defense Program 1961-1969, New York: Harper and Row, 1971.

Hadley, A.T., The Nations Safety and Arms Control, New York: Viking Press, 1961.

Halperin, Morton, Limited War in the Nuclear Age, New York: John Wiley, 1963.

Hitch, Charles J., Decision-making for Defense, Berkeley and Los Angeles, University of Calif. Press, 1966.

Holst, Johan J. & Schneider Jr., William, Why ABM? A Collection of Papers by staff and fellow members of Hudson Institute, Pergamon Press, 1969.

Horelick, Arnold L., Perspectives on the Study of Comparative Military Doctrines, P-5026, May, 1973.

Huntington, Samuel P., The Common Defense: Strategic Problems in National Politics, New York: Columbia University Press, 1961.

Kaufmann, William W., The McNamara Strategy, New York: Harper and Row, 1964.

Military Policy and National Security, Princeton: Princeton University Press, 1956.

Kennan, George F., American Diplomacy 1900-1950, London: Secker and Warburg, 1952.

Realities of American Foreign Policy, Princeton, Princeton University Press, 1954.

Kissinger, Henry A., The Necessity for Choice, New York: Harper and Row, 1960.

The Troubled Partnership: A Re-appraisal of the Atlantic Alliance, New York, McGraw-Hill, 1965.

Nuclear Weapons and Foreign Policy, New York: Harper and Row, 1957.

Kinter, William R. & Pfaltzgraff, Robert L., (eds.) SALT Implications for Arms Control in the 1970s, Univeristy of Pittsburg Press, 1973.

Kaplan, Morton A. (ed.) SALT: Problems and Prospects, University of Chicago, 1973.

Kolodziej, Edward A., The Uncommon Defense and Congress, 1945-1963, Columbia: Ohio State University Press, 1966.

Liska, George, Imperial America, Baltimore, 1967.

Lecht, L.A., Goals, Priorities, and Dollars, New York Free Press 1966.

McNamara, Robert S., The Essence of Security, Reflections in Office, New York: Harper & Row, 1968.

Osgood, Robert E., The Entangling Alliance, Chicago: University of Chicago Press.

Quade, Edward, S., ed., Analysis for Military Decisions, Chicago: Rand, McNally & Co., 1966.

Quester, George H., Nuclear Diplomacy. The First Twenty-Five Years, Dunellen, 1970.

Rosencrance, Richard, Strategic Deterrence Reconsidered, Adelphi Papers, 116, Spring 1975.

Schelling, Thomas C., The Strategy of Conflict, a Galaxy Book.

Schilling, Warner R., Hammond, Paul Y. & Snyder, Glenn H., Strategy, Politics & Defense Budgets, New York: Columbia University Press, 1962.

Schlesinger, Arthur M., Jr., A Thousand Days, Houghton Mifflin, 1965.

Schlesinger, James R., Selected Papers on National Security 1964-1968, RAND P-5284.

Slocombe, Walter, The Political Implications of Strategic Parity, Adelphi Papers 77, May 1971.

Sorensen, Theodore C., Kennedy, Harper & Row, 1965.

Wolf, Charles, Jr., Perceptions of the Military Balance: Models and Anecdotes, RAND P-5402, March 1974.

Wolfers, Arnold, Discord and Collaboration: Essays on International Politics, Baltimore: John Hopkins Press, 1962.

Articles

Bailey, Martin J., "Deterrence, Assured Destruction, and Defense," Orbis, Vol. XVI, No. 3, Fall, 1972.

Beecher, W. "Major War Plans are being Revised by White House", New York Times, Aug. 5, 1972.

Brown, George S., General, USAF, "Technology: The Mould for Future Strategy", Strategic Review,

Spring, 1973.

Carl, Richard L., "Strategic Doctrine in the Nuclear Age", Strategic Review, Winter, 1975.

Carter, Barry, "Nuclear Strategy and Nuclear Weapons", Scientific American, May, 1974, Vol. 230, No. 5.

Cleave, William R. Van & Barnett, Roger W., "Strategic Adaptability", Orbis, Fall 1974, No. 3, Vol. XVIII.

Coffey, J.I., "Strategic Superiority, Deterrence and Arms Control", Orbis, Winter 1970, Vol. VIII, No. 4.

Garnett, J.C., "The Analysis of Threats", International Relations, Nov., 1971.

Gray, Colin S., "Unsafe at any Speed. A Critique of 'Stable" Deterrence Doctrine", June 1973, RUSI Journal, Vol. CXVIII.

"The Urge to Compete: Rationales for Arms Racing", World Politics, Vol. XXVI, January, 1974, No. 2.

"Foreign Policy and the Strategic Balance," Orbis, Fall 1974, No. 3, Vol. XVIII.

"Of bargaining Chips and Building Blocks: Arms Control and Defense Policy", International Journal, Spring 1973, Vol. XXVIII, No. 2.

"Arms Control and Defense Policy", International Journal, Spring 1973, Vol. XXVIII, No. 2.

"Rethinking Nuclear Strategy", Orbis, Winter 1974, Vol. XVII, No. 4.

Greenwood, Ted and Nacht, Michael L., "The New Nuclear Debate: Sense or Nonsense?" Foreign Affairs, July 1974.

Hammond, Paul Y., "A Functional Analysis of Defense Department Decision-making in the McNamara Administration", American Political Science Review

62, March 1968.

Holloway, Bruce K., General USAF, "Reflections on Nuclear Strategy and the Nixon Doctrine", Strategic Review, Spring 1973.

Ikle, Fred Charles, "Can Nuclear Deterrence Last Out the Century?" Foreign Affairs, January, 1973, Vol. 51, No. 2.

Kaplan, Lawrence A., "The USA and the Origins of NATO" Review Politics, 31, 1969.

Kupperman, Robert H. & Behr, Robert M., & Jones, Thomas P., "The Deterrence Continuum", Orbis, Fall 1974, No. 3.

Leitenberg, Milton, "Race to Oblivion", Bulletin of Atomic Scientists, September 1974.

Libby, Ruthven E., Vice-Admiral, "The Role of the Submarine in the Strategic Balance", Strategic Review, Spring 1973.

Martin, Lawrence, "Changes in American Strategic Doctrine--an Initial Interpretation", Survival, July/August, 1974.

Metcalf, Arthur G.B., "SALT II--Some Principles", Strategic Review, Summer 1973.

Rathjens, G.W., "Flexible Response Options," Orbis Vol. XIII, Fall 1974, No. 3.

Schlesinger, James R., "Systems Analysis and the Political Process," RAND Paper 3464, Santa Monica, CA, RAND 1967.

Schwartz, Charles, "Helping the Pentagon Aim Right", Bulletin of Atomic Scientists, Nov. 1974.

Scowcroft, Brent, "Deterrence and Strategic Superiority", Orbis, Vol. XIII, No. 2, Summer 1969.

Seabury, Paul, "On Detente", Survey, Spring 1973, Vol. 19, No. 2.

Snyder, J.D., "Threat Perception and the Armament-Tension Dilemma", _Conflict Resolution_, Vol. II, No. 1.

Sprout, Harold, & Sprout, Margaret, "National Priorities: Demands, Resources, Dilemmas", _World Politics_, 24, 1971-1972.

Westervelt, Donald R., "The Essence of Armed Futility", _Orbis_, Vol. XVIII, Fall 1975, No. 3.